# PYRAMIDS OF LIFE

'What immortal hand or eye
Dare frame thy fearful symmetry?'
William Blake

# PYRAMIDS OF LIFE

Illuminations of Nature's Fearful Symmetry

JOHN READER

HARVEY CROZE

J. B. LIPPINCOTT COMPANY
Philadelphia and New York

Conception, photography and design
by John Reader
Structure and text by Harvey Croze
Photographs on page 185 (full page colobus
monkey) and on top of page 77 (bushbuck 'flash'
opposite dik-dik) are by Mark Reader ©

**U.S. Library of Congress Cataloging in Publication Data**

Reader, John.
Pyramids of life: Illuminations of nature's fearful symmetry

Bibliography: p.
Includes index.
1. Zoology—Africa, East—Ecology. 2. Ecology—Africa, East. I. Croze, Harvey, joint author. II. Title.
QH195.A23R4 591.5'0967 76–50638
ISBN-0-397-01151-2

for Brigitte and Nani

# Foreword

**by Niko Tinbergen**

In the last decade numerous richly illustrated books have been published about tropical, in particular East African, wildlife. Some of them are solid reports about major research projects, and tell us of fascinating discoveries about single species—almost all of them large mammals: chimpanzees, hyenas, lions, elephants. Other books, in fact the vast majority, are primarily picture books that give vivid and often beautiful impressions of tropical scenery, flowers and animals. Most of them are essentially visual displays; they concentrate on the more flashy or the most impressive members of exotic floras and faunas, but they give little information about them. Their function is to delight, to whet the appetite for a safari, and to promote the cause of conservation. And indeed the best among them never pall.

'Pyramids of Life' is neither a scientific monograph nor a 'safari spectacular'. The photographs by John Reader are as beautiful as any (and many of them are real gems), but quite a number of them have been selected because they also have a story to tell. What these stories are is made clear in the text, written by Dr. Harvey Croze, a zoologist who has worked for years in Kenya and Tanzania. His main aim is to give us, in a large series of double-page 'spreads' (in each of which text and figures form an integrated whole), a number of glimpses of the numerous inter-relationships that exist, everywhere, between soil, climate, plants and animals. Of course, Dr. Croze does not pretend to offer anything like an exhaustive treatise; his intention is to alert his readers to a way of looking at organisms in their habitats, in particular to see them as members of vast and complex 'ecosystems'.

The stories are arranged according to one of the central concepts, perhaps *the* central principle of ecology: the circular movement of energy through each ecosystem. The soil, receiving sunlight, water and air, yields the primary producers or plants; these are eaten by herbivores; and these in turn end up largely in the bodies of carnivores (the largest of which are at the very top of the 'Pyramid of Life'). At any point in this ascent from plant to top-carnivore, organisms can die, and their bodies are broken down by decomposers (as were their faeces—natural manure—during their lifetime). Decomposers are creatures, largely microscopic, that transform the complex chemical compounds of the bodies of the higher organisms into simpler substances, which, returned to the soil, are thus once more made available for the start of a new cycle.

The authors make most of their probes into these cyclical processes in the grassland ecosystems of the great plains (section 1); in sections 2 and 3 they pay (rather more fleeting) visits to lakes and streams, and to forests respectively. However rich the variety of life-forms in these and other habitats, none of them is totally separate from others; each of the species is part of one or more of these energy cycles. Yet within this generally valid scheme, each species has its own way of taking up energy and of using and expending it.

The ecological slant of this book can make it an eye-opener for those visitors to National Parks or other undisturbed regions who, though fascinated by the spectacular external features of tropical wildlife, would like to know more about the ways in which all these animals manage to earn a living, and even to co-exist on what would seem to be resources coveted by too many. The text, extremely readable, throws light on such diverse subjects as the causes of the alkalinity of so many lakes in the Rift Valley; the relation between a gigantic eruption of the Ngorongoro volcano long ago and the lack of trees on the Serengeti Plains; the long-term fluctuations of the water level of some of the East African lakes and their influence on, for instance, the fever trees; the curious relationships between elephants and acacia trees; the different and therefore, after all, non-competitive ways in which the grazing herbivores of the vast plains utilise the different kinds, and even different parts of grasses; and numerous other topics. Time and again the book calls attention to the 'system in the madness' of the seemingly endless variety of life-forms and of life-styles.

But the full story is even more complicated. These systems, though stable in the short-term, do vary in time. Dr. Croze has some interesting things to say on these variations. He points out that there are fundamentally different kinds of change. On the one extreme there are long-term oscillations, for instance in climate and therefore in flora and fauna, of which we see only short sections in our lifetime, which we all too easily take for signs of consistent trends. On the other there are short-term changes, of which some at least can be seen to be reversible in a relatively short time. Many of these are due to climatic oscillations, and do not need corrective action by Man. But all too often changes, slow to start with, but rapidly accelerating, are the consequence of human interference, either with the habitat, or directly with animal species. Thus elephant and rhino are at the moment, for different reasons, in very real danger.

While it seems wise to let Nature take her course with her long-term fluctuations, and to adjust to her instead of trying to manage her, the short-term changes induced by Man have often to be dealt with by conservation measures. It is ecological studies which will have to make clear why, for instance, changes in the habitat can be fatal to some species, or why the regional extinction (or its opposite, a population explosion) of one species may have far-reaching, often disastrous consequences for others, or even for the primary plant cover. And it is this kind of knowledge and understanding that those who manage nature reserves want the ecologist to provide. Unfortunately, the complexity of tropical ecosystems makes their research extremely time-consuming, and time is fast running out. National Parks and other uncultivated areas are coming, quite naturally and inevitably, under increasing pressure from local populations, who begin to realise that you can't eat the money that is brought in by safari-going tourists. What these visitors come to admire, and would like to preserve, is to hungry men on the spot above all tempting meat on the hoof. It is not least because of the urgent need to find a way for co-existence of Man and wildlife that the 'ecosystems approach' and attention for energy-flow problems are being applied by a growing number of researchers. May this book contribute to a greater awareness of what is really going on in 'the Pyramids of Life', and so to effective management of what little is left of the natural riches of Africa, and indeed the Tropics generally.

N. T.

# Contents

# Introduction

There is a fearful symmetry in the natural world, an overwhelming order we are just beginning to understand. Some might argue that nature is best appreciated unfettered by human explanations; best enjoyed as a relief from the man-made world. But we have come too far for that luxury.

Man's ravages of nature have been well enough chronicled in recent years. But too often the emphasis has been on the sensational elements of extinction and destruction; too often we have begun to tackle the problems of the sick and dying before fully understanding the functional success of the healthy and living.

Just as it would be foolish to tinker with a complicated and valuable watch without first understanding what makes it tick, so we must learn how the natural world functions if we are to keep it – and thereby ourselves – in good working order.

Our first response to anything in nature is usually concerned with whether or not it 'looks nice'. An emotional response, and one that is particularly evident when we first visit tropical environments and are confronted with their bewildering array of remarkable objects and events, many of which we may never have seen before. We delight in the beauty of a butterfly, and shudder with revulsion at the ugly vulture. The butterfly is immediately acceptable, we may even want to know more about it; but the vulture, with its looks and predilection for corpses, is more likely to be dismissed simply on the strength of its appearance.

These prejudices potentially close our minds to important aspects of life, and the emotional concepts of beauty and ugliness are actually more of a hindrance than a help to our understanding of the natural world. We must take a broader view. This is not to say we must all understand the science of entomology to appreciate the beauty of the butterfly, but undoubtedly the vulture's ugliness will cease deterring our interests when we understand how essential the creature is to the functioning of the tropical grasslands, and how essential its appearance is to its own functioning.

Ecology is the science which deals with the relationship between living organisms and their environment. The ecologist shares our wonder when viewing the beauties of nature; he is a naturalist at heart and his first reactions are often no more complicated than our own. But then his curiosity gets the better of him, and he goes on to observe life's patterns closely, and to organise his knowledge of them.

We cannot all be ecologists, but nevertheless, our understanding and therefore our appreciation, of how the natural world works will flourish once we begin to look at its objects and events in their ecological perspective. The foremost principle of this perspective is what might be called the Rule of Non-Randomness: there is nothing superfluous in the natural world. The living earth is made up of biological systems, it is not simply a haphazard collection of objects and events. A forest is an arrangement of trees. Patterns, sequences and interactions occur in nature with the regularity of streets, traffic flow, shop distribution and electricity use in a large city.

This means that when you look at a butterfly in a particular place, or a vulture swooping down onto the carcass of a gazelle, you are not

watching random isolated events. You are invariably glimpsing part of a system, part of a biological process that is both measurable and predictable. Given just a little more knowledge and experience, the mere presence of a butterfly could tell you something about the vegetation, the soil and climatic aspects of the system it is in. And the condition of the gazelle carcass, the species of the vultures about and their numbers, could tell you quite a lot about the system's prey and predator populations.

Of course, in a global sense, the earth itself is one vast biological system; but within it we can define thousands and millions of quite distinct systems; each a self-contained unit made up of a finite collection of the fundamental elements of life. The system could be a pond two feet across, or 2,000 square miles of grassland – whatever the size, the principles governing the way it functions remain the same.

You could think of a biological system as a complex machine, built with the elements of a living system – climate, soils, vegetation, and animals. These elements are constantly shifting their relative positions; but they always remain very closely connected by common biological laws, by their effects upon one another, and by their dependence upon one another. Consequently, natural events form patterns we find repeated in all biological systems, even in those that look completely different.

This fundamental fact has determined the form of our book, which is divided into three parts: the Grasslands, the Lakes and Rivers, and the Forests of tropical Africa. In each part we follow the cycle of nutrients, from the earth to the vegetation to the animals and back to the earth again. Along the way we demonstrate how just a few of the earth's countless life-forms deal with the three basic facts of life common to all: how they find their food; how they avoid becoming some other organism's food; and how they perpetuate their species. The habitats to which our characters belong range from bare rock to forest; from soda lake to semi desert. The life-forms themselves include blades of grass, trees, slime moulds and lichens, microscopic protozoa and elephants, insects and reptiles, cats and a frog or two.

And our purpose? It is to show that even amid such an apparent confusion of place, object and event there is always order; the very diversity itself is determined by biological laws that no organism can escape. And understanding this, the symmetry of the natural world is no longer fearful; it is the beauty of it.

We start our investigation into the underlying order of the natural world with life's building blocks, the chemical elements of the earth itself. Every once in a while the earth reminds us that she has even yet not finished forming: a city crumbles or a new island appears. Oldoinyo Lengai, the Mountain of God in northern Tanzania (illustrated on the title page), is only a few million years old. It last puffed ash over the Serengeti Plains in 1966. The materials bound in its barren volcanic mass are the potential exchangeable mineral resources to feed a living system, but first they must be freed from the rock. Slowly, slowly, sun and rain, heat and cold, organic acids and the roots of pioneer plants insinuate themselves into the physical and chemical chinks of the parent material, which crumbles, mixes with the organic débris and becomes soil. Unbound, the mineral resources join the roundabout of life.

Consider one atom of one element in the soil, say phosphorus–a particle of one of the usable mineral resources in a living system. If we could tag that particle, by making it radioactive perhaps, we might be able to trace its movements through the system. The atom is absorbed by the root of a plant and is incorporated into the plant itself, perhaps as part of a cell wall. If the plant is a grass, it might be eaten by a wildebeest. Our atom of phosphorus is then perhaps used in building wildebeest flesh. The wildebeest might be caught and eaten by a wild dog, and the same atom of phosphorus which was once part of the soil, now becomes part of the wild dog. The dog excretes, or dies, and the atom is thus made available to the maintenance of yet another class or organism, the decomposers, those animals from bacteria to vultures who complete the disassembly of organic structure, who bridge the gap between life and death and thus return our atom of phosphorus to the soil. It is then ready to be run through the system once more.

The path the atom has taken is called a food chain. There are alternative paths in all biological systems called food webs, interconnected food chains, complex alternative paths by which mineral resources move through a system. The same grass could have been eaten by a gazelle and the phosphorus excreted out the next day, or it could have passed into a cheetah. Thence to a vulture when the cheetah dies, or to other decomposers when the vulture excretes and back, inevitably, to the soil.

Having followed the path of our element from the soil through the plant and animal communities back to the soil, we would find ourselves not very far from where we started – probably not more than one hundred miles, usually less. We would have been travelling within an ecosystem, an integrated, self-maintaining biological unit. An ecosystem is the set of all the places through which a particular atom is likely to pass. It is therefore a system of all the soil types, plant and animals in a particular area.

Every ecosystem has its own special characteristics; it has for example, a geographical location. The boundary of the ecosystem often coincides with a number of physical boundaries, like a Rift Valley wall, or a lake shore, or the interface between two soil types; when it does, we might find vegetational boundaries congruent with the physical ones, such as the line between forest and grassland. As we shall see later, most animals usually stay on one side or the other of ecosystem boundaries. Of course, boundaries need not be barriers: it is possible for animals to move across them. But the interdependencies between soil, plants and animals make it more likely that the atom of phosphorus will stay within the ecosystem.

Different ecosystems have different weights – or biomasses – of plants and animals, usually a function of the climate and the amount of nutrients available from the carth. And ecosystems look different, too, because of their different topography, soils and plant formations.

There must be, of course, some form of energy that moves our atom and all the other materials round and round an ecosystem. The sun is the only natural source of energy in our solar system; its radiance gives light and keeps us warm. But no organism can eat sunlight. The only way the sun's energy can be trapped is in the process of photosynthesis by green plants, that is to say 'making with light'. What do plants make?

Given the energy of the sun acting on water from the earth, carbon dioxide from the air, and the wonder ingredient chlorophyll in the plant cells, green plants produce oxygen and simple sugars – the first edible collection of energy and materials in the food chain. This is the primary production of a living system, the basic 'trophic' or nourishment level. Thereafter, the sun's energy, alternately trapped and released in certain chemical bonds, pumps through food chains and travels with the minerals from one trophic level up to the next: from *Primary production* to *Herbivores* (eaters of plants), to *Carnivores* (eaters of flesh), to *Decomposers* (eaters of everything).

All the energy that enters our ecosystems eventually goes out again, dispersed into outer space in the form of heat. Turn off the sun, energy ceases to flow through the finite collection of materials and the system runs down, as in a car that has run out of petrol.

The Second Law of Thermodynamics tells us about the behaviour of energy in a system. The law states that if you change the form of energy you lose some of it in the process, as heat. You get hot when you run, because of the heat produced when the chemical energy of your food supply is changed into the kinetic energy of movement. Just growing or replacing worn cells requires energy conversion, so all organisms, even blades of grass, give off heat, the tax of living.

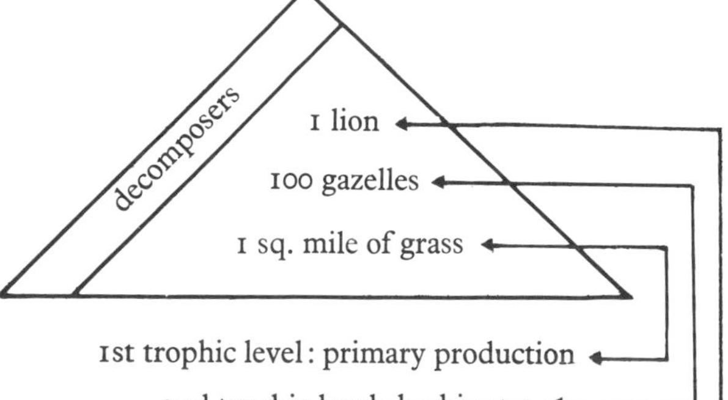


Grass is only concerned with being grass, so most of the energy it absorbs from the sun is dispersed in heat as the grass maintains itself. In fact only about one tenth of the sun's energy which became grass remains to be eaten by, say, a gazelle. Gazelles must maintain themselves too, and in doing so use all but one tenth of the energy they derived from the grass. Hence the leopard that eats the gazelle gets in the end only one hundredth of the sun's energy which was originally trapped by the grass.

The Second Law of Thermodynamics thus explains why there could never be more gazelle meat than grass tissue and never more leopards or lions than gazelles: there is not enough energy to feed them. The mandatory heat loss as each organism in an ecosystem maintains itself decides the relative amounts of energy, numbers and biomass in the trophic level above. The energy inevitably diminishes as we move up through the food chains and a pyramid of life is formed which applies to all ecosystems.

The decomposers, we see, are a special class of beasts who eventually feed on all trophic levels. Nevertheless, they too are subject to the law, and must exist at a smaller biomass than their food source.

The primary production is the *basic food source* in every ecosystem. Each habitat presents its basic food source in a particular pattern, the form and extent of which is determined by geology and climate (the *controlling factors*), and modified by fire and by the animals eating it (the *modifying factors*).

When we talk of eating and of the movements of energy and materials from one trophic level to another, we begin to cross the thin line between the sciences of ecology and behaviour. Materials do not move passively, like water running from one tank to another. Materials are moved by animals behaving, by their *pursuit of food*. When a lion eats a gazelle, the elements in that gazelle at once change their position in the pyramid of life. Of course, the gazelle will try not to become the lion's food and it is by *avoiding predators* – the obverse of feeding behaviour – that a species attempts to keep its materials in its own trophic level. We can picture these two sorts of behaviour as taking place at the boundaries between trophic levels.

Within a trophic level, we observe in *reproduction* the complex rituals and elaborate adaptations of breeding behaviour. These maintain the integrity of a species, and the internal organisation of an animal or plant community. They are part of species' *social organisation*, which quite apart from ensuring breeding success, may also assist in the pursuit of food, or in predator avoidance. All such behaviours keep materials moving around within a trophic level, rather than from one to another. Throughout the book the headings indicate which basic features of the ecological pattern – italicised in the paragraphs above – are being illustrated.

We have said there is nothing superfluous in the natural world; so there must be a reason for everything, an answer to every question: Why is an organism built the way it is? Why does a plant grow in a particular place? Why does an animal behave the way it does? We shall try to answer these kinds of questions, though we do not presume to give definitive answers to all we raise – many answers are not yet known. But in any event we will try to show how a curious naturalist might think about the question and set about finding the answer.

In one sense there are as many answers as there are organisms, answers that have to do with place and time, but these are only part of the basic answer, which is really quite simple: an organism is the way it is, and does what it does, in order to stay alive and well long enough to reproduce more of its kind. We might call this the Rule of the Ultimate Function, for indeed it leads us to the most important of life's concepts that should be understood at the outset.

An animal or plant is not a machine: the complex of chemical information encoded in the genes in the heart of every cell cannot be perfectly replicated each time an individual reproduces. Chance differences are bound to occur. For this reason we find variability within species, or even between siblings. Some products of reproduction fare better, others worse. The judge in the lottery of breeding and survival is the environment. Those organisms which are better adapted to life in their surroundings will produce more young, bearing the very characteristics which gave their parents a survival edge. In this way, the environment is said to select the best strategy for survival from the number of alternatives offered each time there is reproduction and a recombination of the genetic information which defines a species. We have, then, Natural Selection, the most important concept of life in the natural world – the one which dictates that order will exist.

# I

# GRASSLANDS

## Primary production: Controlling factors

# Geology and Climate

The African grasslands support the highest concentrations of large mammals in the world. These habitats are so extensive that if we were to walk from the Sahara to the Cape we would trudge most of the way through various types of grassland. The rolling savannah dotted with flat-topped acacia trees is part of the picture of Africa everyone knows. To the rangeland ecologist it is known as **'wooded grassland'**. If the tree canopies cover more than about twenty per cent of the ground, the habitat grades into 'woodland'; where there are more bushes than trees, it is 'bushed grassland'; the nearly treeless areas are simply 'grasslands' or 'plains'. What controls the existence of these extremely productive habitats?

Their extent is determined by a blend of altitude (2,000 to 6,000 feet); undulating topography; an equitable temperature (50 to 80 degrees F.) despite intense solar radiation; poor and sporadic rainfall (20 to 30 inches annually) distributed in distinct seasons; and shallow soils, often red on the slopes, black clay in the sumps.

Soil differences may be just local changes down a hillside or may be very widespread. For instance, the ashes blasted from Ngorongoro crater when it erupted five million years ago fell and covered 7,500 square miles to the north and west. The resultant ash-rich alkaline soils inhibit the growth of trees, and the soil boundary therefore marks the edge of the virtually treeless Serengeti Plains. Beyond the plains where the soil is kinder to their needs, trees and shrubs are able to grow again.

The grasslands, then, are a mosaic of different forms of vegetation controlled by local differences in topography, soils and rainfall. These inanimate patterns determine the animate patterns we encounter in the grasslands. In this picture for example, thick riverine bush shows clearly the drainage patterns of the area because its growth along those particular lines is encouraged by the additional rain water retained in the dips and depressions of the earth's surface by the fine soils which have been washed down from the higher slopes.

# Rain

Seasonal rainfall distribution can have dramatic effects on grassland vegetation patterns. The difference between the dry and wet seasons has to be seen to be believed: it is often the difference between a lush meadow and a dustbowl. Yet the rainfall in the grasslands is not only low, it is also too unpredictable to produce the kind of steady-state vegetation found in temperate regions. Seventy years in every hundred will receive less than thirty inches. And the rain that falls is not evenly distributed for the convenience of plant reproductive cycles. Within the year, most rain crashes down in a couple of months – the wet season; or else in two relatively brief periods – the long, and the short rains. Furthermore, it falls too hard and runs off too quickly for the soil, and hence the plants, to absorb it all. The rain which does soak in is just enough to sustain those forms of life adapted to survive through the eight months of the dry season. The effect on vegetation of the annual rainfall distribution becomes strikingly clear when we consider that the wooded grasslands of Africa receive the same average precipitation as Ireland, the Emerald Isle, where the rainy season is continuous and the evaporation far less.

Again, quite apart from these annual fluctuations, the rain falls irregularly on the larger scale of decades too: either coming, for instance, in the flood-producing torrents which East Africa experienced in 1961; or failing entirely, as during the region's 1971–72 drought; or falling 'normally', as in the intervening years. To survive these moisture extremes, plants have developed and maintain what might be called an ecological agility – an ability to recover remarkably quickly after long periods of drought. They lie dormant for months, seemingly dead; but then, within days of the first rain, they sprout; soon trees are green and flowering, land that seemed certain to become barren desert is transformed into productive pastures for the **wildebeeste** and the numerous other herbivores. Some trees even begin to grow before the rains arrive, cued by signals to which we are insensible.

Once rain has fallen, the amount ultimately available to the plants is largely dictated by the inevitable re-appearance of the sun after the downpour. For the sun, which on the one hand provides energy for the prolific tropical primary production, takes its toll on the other by evaporating nearly eighty per cent of the rainfall, as well as sucking moisture from the leaves themselves. If the climate were to become drier, the grasslands would tend to become desert. If rainfall were to increase, trees would get enough water to grow in number and literally overshadow the grasses: woodland or forest would result.

The grasslands may go, but grass of some sort will always persist, though ultimately dependent, like most living things, on water. Rain is inanimate, so to call it 'life-giving' is not quite correct. But undeniably, once the earth has set the stage, the seasonal showers that sweep across the land are the cue for all the acts we observe in the grassland, or indeed anywhere else.

# Grass

Grass is the single most successful visible terrestrial life form, both for propensity to persist and its capacity to increase. Virtually anywhere on earth, we will not walk far before treading on grass.

A relatively large part of the biomass of grassland plant communities goes into the production of offspring, seeds. In the form and structure of a grass plant the emphasis on the seeds is particularly striking – a few thin leaves, one or two stems and a seed head which may weigh as much as the rest of the plant. Clearly these are organisms whose success lies in their ability to flourish when conditions are right. Since the correct conditions may be limited to a few short weeks of rain in the tropical grasslands, the grasses there have evolved to grow and reproduce as quickly as possible. One might call them 'increase strategists', particularly the fast maturing annual grasses which grow, reproduce and die in one short season.

Organisms which produce an abundance of seed are ideal colonisers, especially when the seeds are equipped with devices to enhance dispersal – special hooks to attach to the hair of a passing animal, tasty packaging and an impervious seed coating to attract herbivores on the one hand and survive their digestive juices on the other, feathery appendages to catch and be carried by the wind. If conditions are not favourable for flowering, the plants propagate by vegetative means, literally creeping and rooting over the bare surface.

Many grass species are able to grow where nothing else can. In volcanic grit which attracts and holds solar radiation because of its blackness, the surface temperature may exceed 110°F. Even when it rains, water is only available to the roots for a very short time because of the porosity of the top layer. Roots of many species have special microscopic hairs which can absorb the minute moisture droplets which condense at night underground. Annual grasses characteristically have deep root systems which seek out water down to the parent rock.

Nutrients may be scarce if the soil is in its infancy like this. Yet pioneer grasses can make do on the very margin of life support. They continue to break down soil particles with their searching roots, and add to the organic content of the new soil when they die. They therefore modify their habitat, change it so that eventually it becomes suitable for successive forms and at the same time, unsuitable for themselves. But this does not matter; they can move on. The continually changing surface of the earth in the form of a lava flow, a hippo wallow or a roadside verge will always provide a home for the plant pioneers.

On more mature soils where the opportunistic lifestyle of the annual grasses is not called for, perennial grasses take over. We call them 'equilibrium strategists', for they maintain themselves persistently from season to season, at a high population density. Their root systems are tenacious near the surface, and capable of storing food for the dry season. They only tolerate a few deep rooted wild flowers, trees and shrubs, and then only if the local rainfall is sufficient to provide a surplus after the grasses have absorbed all they need at the surface.

**Primary production:**
**Basic food sources**

# Prolific Pastures

The primary production of the grasslands is prolific; during the rains, for example, every square yard of grass can produce a pound of edible material each month – 1,500 tons to the square mile. This rapid conversion of materials into an edible, available form creates the opportunity for numerous herbivores to exist. In terms of pounds of large animal flesh per unit area (i.e. biomass density), the Serengeti National Park in Tanzania and the Ruwenzori National Park in Uganda each support something like 200 times as much as the forests we shall look at later. In the **Serengeti**, the live-weight of the 26 species of herbivore, from elephants to dik-diks, is in the region of 1,000,000,000 pounds – roughly the same as the weight of the inhabitants of Greater London. And that does not include the 10 species of primates, 27 species of carnivore, more than 500 species of birds and the uncounted thousands of invertebrate species – which themselves probably weigh as much as all the vertebrates put together. The animal biomass of the wooded grassland is unequalled in any other living system.

For the plants, production and survival is determined largely in the dimension of time: water for growth and reproduction is only available in the rainy season. What plants do in time, animals and people do in space. The migratory herds and nomadic pastoral tribes of Africa have evolved an opportunistic way of life; for them, survival lies in continually moving from the dry and withered to the wet and green. And if geology and topography, rainfall and evaporation are the factors controlling or limiting the form of the vegetation, then it is the moveable animal feasts, combined with fire, that modify it. Such interactions – between flora and fauna, between the animate and the inanimate, between fire and water – keep the cycles of the natural world in motion.

**Primary production: Modifying factors**

# Fire

In primeval times it was a freak lightning strike or a spark from a volcano that started grass fires. Today it is more likely to be a man with a match. But, whatever the cause, fire has become an integral modifying factor of grassland ecology – until recently over three-quarters of the Serengeti National Park burned each year. A few tree species have adapted to withstand burning by developing a thick fireproof bark, others 're-treat' underground: a stunted but living two foot high acacia in a regularly burned area might have a root stock six inches thick – evidence that the part of the plant above ground has been burned back regularly for a decade or more.

Some grasslands are maintained by fire: others modified as fire-resistant shrubs creep in. The effects are varied, depending on local conditions.

It could be argued that fires are the quickest way to break down and return materials to the soil. This may be true, but there are several disadvantages. For example, the materials which were grass will not always fall back to the soil from which they grew.

In the heat of a dry grass fire, ash and smoke rise thousands of feet; blown on the wind they spread a delicate blue haze as far as the eye can see. It produces spectacular sunsets, but represents the exportation of tons of materials from the grassland ecosystem. A volatile element like nitrogen is lost into the atmosphere in prodigious amounts. Another disadvantage is the cost to other organisms. For herbivores, the food supply disappears in a flash. In fact so do most of the small mammals, if the fire is a major one.

The green shoots which appear after a burn are the reason pastoral man lights grass fires. But the benefits may be only short term. If the fire was lit just as the grass was setting seed or just before returning nutrients to the roots for the dry season, the toll on the grass population could be disastrous.

If we put aside preconceptions, born in temperature climes, that lead us to believe all vegetation undergoes successive stages of growth until a climax form is reached, then we are better able to understand the frequently dramatic changes in Africa's wooded grasslands. In many areas trees are dying and woodlands seem to be disappearing. Is this an ecological disaster? Perhaps not, if the wooded grassland is not a climax form, not an end product in itself, but simply an integral part of a much larger – perhaps centuries long – cycle.

Consider a simplified hypothetical example of a woodland – grassland – woodland cycle of the kind which is typical of the Serengeti today. Old trees begin to die, burned perhaps into grotesque sculptures (left), or pushed over by elephants. Their death allows seedlings which would otherwise have died in the shade of the adult canopy to mature. At the same time more robust species of grass can now thrive. But, more grass combined with a 'bad' dry season is a fire hazard, and should fire occur, regeneration of the acacia woodland is retarded as the young trees are burnt back. This allows still more grass to flourish which, combined with a series of 'good' wet seasons, produces more food for the grazers like wildebeeste and gazelles. Increase a population's food supply and it will, to a point, increase in size. More grazers eat more grass, thus reducing potential fuel for a fire. Young trees begin to grow beyond the fire-critical height of ten feet. Grassland begins to revert to woodland. And so on. It may take a century or more.

Maasai in Amboseli remember through verbal traditions how the yellow-barked fever trees (*Acacia xanthophloea*) all but disappeared some ninety years ago, just as they have been doing over the past ten years.

So, we see that our pyramid of life is a slightly simplistic representation of the interaction between herbivores and plants, for as we examine our basic theme of the flow of energy and resources, we find numerous variations – different habitats composed of different plants and supporting different animals. We need a deep understanding of the effects of plants and animals upon one another before we can predict the patterns. Nevertheless, the shape of the pyramid of life remains fixed in this part of the world where the easiest things to predict are the intensity of the sun and the irregularity of the rains.

**Primary production: Modifying factors**

# Animal Furniture

Animals, as we shall see, are capable of modifying the vegetation of a habitat. They can also have a considerable impact on its physical structure. Of course, over-use of the vegetation may increase the rate at which rain removes the soil and thus lead to erosion. But in undisturbed animal and plant communities, such serious modification rarely, if ever, occurs.

Mud wallows are an important type of 'animal furniture'. They are, like our furniture, literally made for comfort, and are as inoffensive to the habitat as an easy-chair is to our living room. The rolling and dusting of a seasonal migration of zebras creates a bared, slightly depressed opening in the grass cover. Elephants pause to suck up and blow clouds of dust over their backs and thus increase the depression's depth. During the next rains, water accumulates, and with the onset of the dry season a muddy patch is left. A family of warthogs may spend the heat of several days rooting and rolling in the mud. They are displaced by some bull buffaloes who further churn up the mud and carry off more dirt when they leave to feed.

The elephants pass by again and plaster mud on their backs. The next rains produce a pool which persists even further into the following dry season. After several years of use and modification by a whole string of animals, the result is a mud bath capable of entertaining a dozen **hippopotami** from a nearby lake.

All these herbivores cool themselves and escape from worrisome insects in the mud. They also tend to take the pause in the hot day as a natural time to excrete. Water, mud, and the presence of the dung seem to encourage other animals to excrete, so the wallow eventually becomes a pool of concentrated nutrients. There comes a point, perhaps when the wallow gets a bit too rich, when it falls out of favour. The animals begin their furniture building at another spot, and the wallow is left to encroaching plants. Years later, the original wallow is marked only by a patch of particularly lush grass. Grazers still drop by, but now for another reason.

# Animal Feeding

From any point of view, the **elephant's** impact on primary production is as spectacular as its size. In its continuous 'predation' on plants, the elephant tears branches from trees, pulls great tufts of grass and roots from the earth, gouges huge holes in baobabs and pushes over Acacia and Commiphora trees. Is this destruction? Not really; it is modification, perhaps even enrichment of the habitat. For a rich habitat is not one which simply makes a pretty picture, but one in which energy is flowing through numerous pathways, and in which materials are changing form continuously.

In the grasslands of Africa, there are no hard and fast rules which dictate that there will be precisely so many trees and so many elephants. If elephants, or voles for that matter, increase their numbers and reduce their own food supply, they will become undernourished. A proportion of them will die. As numbers drop, the vegetation will begin to change back to a more favourable state.

The wide range of plants which are physically available to elephants as food, their catholic tastes, and their ability to cover thirty miles a day at a comfortable walk, allow them to thrive and leave their mark in nearly all African habitats from semi-arid bushed grassland to mountain forest. The only limiting factor is water, which must be in good supply to support a population whose average member requires fifteen gallons a day. An elephant may spend sixteen hours out of the twenty-four feeding.

Considering that a large elephant can ingest some 300 pounds of green matter a day, we might wonder how the habitat can take it. Yet far from being demolition agents, elephants are in fact the greatest natural construction crew in Africa, contributing more to the change and variety in local habitat than any other species. Nowhere, not even in regions of apparent over-population where habitat destruction might be expected, has it yet been demonstrated that elephants by themselves degrade habitats to the point of creating deserts. They change them certainly, but they do not destroy. When they knock down trees in a woodland for instance, a greater variety of vegetation frequently results – other species of grasses, shrubs, bushes and herbs take hold. In this way elephants increase diversity.

# Legs and mouths

Materials in an ecosystem must move from the plants to the herbivores, from the herbivores to the carnivores, and thence through the decomposers back to the soil. The gateway from one trophic level to the next is through the mouth. The mouth parts of a **harvester ant** (formicidae) are a far cry from those of a **hippopotamus**, yet both animals feed on grass. Like a lawnmower the hippo trims the grass sward with its wide lips, taking in each bite enough grass seeds to keep one harvester ant busy for most of its life.

The mouths of large herbivores characteristically have lips to gather, incisor teeth to clip, molar teeth to grind. Browsers who feed on leaves of shrubs have delicate snouts to select single leaves. Grazers have wide muzzles to harvest grass. Gnawers have pointed faces and tiny mouths, just a bit larger than necessary to expose the working surface of the two front teeth. Such morphological (body form) features are invariably accompanied by characteristic behaviours. Hippos bite from the leaf table, gazelles select single leaves, rodents pick seeds apart, harvester ants cart them away.

There are other necessary adaptations of herbivore body shape and form. The legs of most large herbivores must be long enough to move the animals efficiently over long distances in search of food or for avoidance of predators. This means that necks must be long enough to reach the grass made remote by the legs. Once the head is near the plants, the nose, whiskers and lips must be sensitive enough to select the correct parts of the grass by smell or texture. Add these features together and we have the general shape of a typical antelope – a wildebeest or an impala, for example.

Digestive systems have evolved chemical and mechanical means to unlock the nutrients in cellulose-armoured plant cells. In the process of rumination, the plant material is chewed and fermented several times in succession. Plant cell walls are further broken down by single-celled microorganisms who live in herbivore guts and who produce a special enzyme to digest cellulose. The relationship is symbiotic – advantageous to both: the protozoan gets its food delivered, the herbivore gets its digested.

# Migration

Although plants cannot move, their effective availability changes in time. For the herbivore, grass which has withered away might as well have run away. Thus, as the seasons change, herbivores must move on to greener pastures. The movement can be a relatively modest one, such as that of resident impalas who cluster along rivers in the dry-season and move back up the slopes in the rains – a matter of a mile or two. Or, the movement can be a full scale migration, such as that of the Serengeti's nomadic **wildebeeste** who cover nearly 10,000 square miles in their annual 300 mile round trip of the Serengeti ecosystem. This moving mass, over a million strong, is one of the last great animal spectacles on earth.

The question of why they migrate is both simple and complex. The movement itself serves to keep the population in areas where there is more than 600 pounds of green grass per acre. When drying and eating reduces the amount below this level, the wildebeeste simply move on. Due to a rainfall gradient from the drier south-east to the wetter north-west, the short-grass Serengeti plains begin to dry up first. So around the end of May, the wildebeeste move off to the west and north, through the woodlands, feeding and rutting along the way. The peak of the dry season, July and August, finds them in the still green Kenya Mara. Here they circle around, with the females growing heavy, until November, when the beginning of the short rains greens their path again to the south, where they calve. But what exactly tells the wildebeeste it is time to go, and which way to go?

We hear reports of wildebeeste herds 'deliberately' setting off towards distant rain storms. Do they see the dark clouds and 'understand' the implications, or smell the rain and 'conceive' the link between precipitation and production? Perhaps they have such associations built into their genetic information.

It has been found that the grasses to which the beasts migrate have a higher calcium content than those which they have left. Are they responding to specific mineral 'hungers' which are only satisfied by going elsewhere? Then again, herbivores select grass to eat, not so much according to species of grass, but rather according to its tensile strength. Grass which is easily bitten off is clearly leafier, younger and hence more nourishing. Are the cues then the physical structure of the grass? Such questions will only be answered with careful observation and experiments both with wild and tame wildebeeste.

The phenomena of the wildebeeste migration and of their numbers are controlled by distribution and amounts of primary production. The Serengeti population has been increasing steadily over the past two decades, and is still growing. Part of the reason may be an increase in the amount of open grasslands because of the reduction of woodland by fire and elephants. And part of the population increase may be due to a recovery growth spurt after a devastating epidemic of rinderpest at the end of the last century. Population growth of such magnitude is dramatic but could be expected of any population. The really intriguing question is, when and how will it stop?

## Herbivores: Pursuit of food

# Niches

Different herbivore species, some of whom appear quite similar, use particular parts of the same habitat in different ways. They achieve this by adopting what are called ecological niches, which are similar in concept to human vocations. In the broadest sense, they define everything a particular species does for a living; but a niche is a tricky thing to describe, since the exercise demands such an intimate knowledge of an animal's way of life. It is easiest first to describe where it is.

Herbivore food does not run away from the herbivore, but it is effectively available in different measures and qualities in different places. Specialization on one class of food is, among other things, a way of conserving the energy involved in searching – you look in one place, not everywhere. Hence there are some herbivores who are riverine species: **waterbuck** (a) and reedbuck; those who fill an open grassland niche: gazelle and **wildebeest** (b), **zebra and ostrich** (c); those who fill a wooded bushland niche: kudu and **bushbuck** (d); and those who use the ecotone, the area of intergradation between woodland and grassland: topi and **impala** (e).

An important cause of such diversity of types and ecological niches is competition for common resources. Back in evolutionary time the species were very similar if not, in many cases, one and the same. Niches become 'crowded' as the number of organisms using common resources increases. One way to avoid conflict is for one of the contestants to change its tastes. If the divergence increases its chance of survival, it is on its way to becoming a new species, and over the aeons, will slip into another ecological niche. The effect, of course, is to reduce competition between species, to increase the diversity of the animal community and to increase the number of paths energy can take through the ecosystem.

The **eland** on the horizon above is a plains animal, one of the specialists with a rather narrow niche. It feeds predominantly on dicotyledonous plants, on the leaves of small wild flowers which grow amongst the grass. Such plants are nutritious, so the eland can build a large body. Males can weigh up to a ton and stand six feet at the shoulder. But these plants are scattered and relatively scarce. Hence the density of elands must be low and their distribution scattered. The eland is a good example of the ultimate dependency of the upper on the lower trophic level, of the herbivore upon its food supply.

a

b

c

d

e

# The Grazing Sequence

Niche separation in space and time can be obvious as we have seen – it can, as well, be too subtle for the human observer to discern at first glance. Animals may appear to be sharing the same pasture, but every farmer knows that the horses in his field are eating different things from the cows. Amongst African herbivores, resources may be shared in a mutually beneficial way known as a grazing sequence.

Because of its coarse structure, a mature stand of grass is physically unavailable to the little gazelles and even unpalatable to the wildebeeste. But the heavy feeders like buffaloes and elephants can tramp into the chest-high grass and efficiently harvest what they need. Once they have trampled and eaten the heavier material and permitted some re-growth of finer grass, the zebras and wildebeeste move in. The **zebras** are also, to some extent, coarse-feeders and do not mind grass with a moderately low leaf to stem ratio; they are followed by the **wildebeeste** (middle) who prefer a bit more leaf and less stem. But both are daunted by the virtually inedible, potentially dangerous stands of tall grass which could conceal predators, so they 'let' the larger animals prepare the way.

After the zebras and the wildebeeste have refined the structure of the pasture even further, and mulched some uneaten material into the soil which stimulates the growth of more shoots, the grassland is then 'ready' for the herbivores with more delicate mouths and refined tastes – the hartebeeste and gazelles. These selectively nip off the leafy parts of the grasses or pluck the dicotyledons left by the bigger animals. Nor is the sequence finished. **Egyptian geese** (right) graze on the shortest swards, termites clean up the debris, and insectivorous birds take continual advantage of the insects stirred up by the feet of the herbivores. Thus in a grazing sequence, the same acre of grassland may be used sequentially and simultaneously during the growing season by a number of herbivores. The result is a healthy grass 'lawn', often a mosaic of close-cropped patches of favoured species and longer patches of stemmier grasses.

The primary production under such a régime of use by a mixed species herbivore community, is actually greater than if the grassland were left untouched. Use begets growth, and secondary productivity maintains a maximum level as long as the dynamics and timing of the succession are not interfered with. If one species stays too long or if fire eliminates the grass, the succession is thrown temporarily out of rhythm. The system is flexible enough to weather such disturbances at their natural, infrequent rate. If drought or human activities such as burning or the confinement of the grazing herds, disturb the sequence for too long, then the result is over-grazing, weed encroachment, bare ground and erosion. The grassland begins to die.

a

b

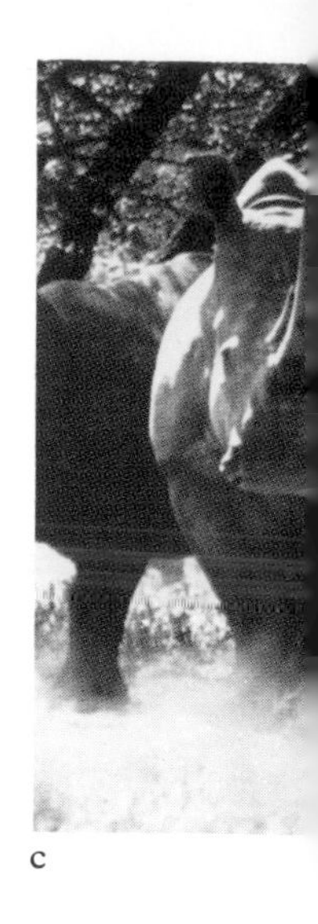
c

e

d

f



# The Evolution of Species

Evolution is often thought of as something which happened ages ago, an event which, like the Creation, is now finished. But life forms are constantly, if slowly, changing in response to the demands of an ever-changing environment. Modification of form and behaviour will go on as long as there is life; species will continue to come and go. The contemporary dynamism of evolution is most apparent when we observe species which look almost the same, but which, on closer examination, turn out to be quite different.

From a distance there is not much to distinguish **Grevy's** (a) and **Burchell's** (b) **zebra**; or the **White** (c) and the **Black** (d) **rhinoceros**; or the **Reticulated** (e) and the **Maasai** (f) **giraffe**. But they are different – the Grevy's stripes are thinner and the ears larger; the larger white rhino has a distinctive squared-off upper lip (hence its name 'white', which is not a description of its colour but a corruption of the Afrikaans word for wide); and the reticulated's pattern is bolder than the Maasai's. Why should they be different at all?

In the first place the similar forms live in different places. For example, the reticulated giraffe is an animal of dry country like northern Kenya and is separated from the nearest Maasai population by one hundred miles. Both must have arisen from a common ancestor and Pleistocene populations separated by a geographic barrier, or by a chance one-way migration, each responded in its own way to the new local conditions. All populations of one species separated in space tend to become distinct 'ecotypes', different forms of one species. The two giraffes are examples.

Isolation results in changes in behaviour as well as colouration or body form. After thousands of years apart, the Grevy's finds that it has a different social system from the Burchell's zebra; a territorial one designed to conserve a scarce grass supply in arid areas. Their behaviour thus diverged enough to isolate effectively the two incipient species. When such populations happen to come together again after many generations of separation, they tend to remain distinct, as populations of the two zebras do which now overlap in northern Kenya. It is doubtful whether they would pass the test of true speciation – if they could be convinced to overcome their behavioural barrier and mate, they would quite likely produce a fertile hybrid. 'Good species' would not.

Both giraffes eat acacia leaves and both zebras eat grass, but the two rhino species have gone further in their speciation. The black rhino, as we have seen, is a browser; but the white, as we would guess from his broad lips, is a grazer. Body form and lifestyle changes of this magnitude suggest that the two rhinos diverged a very long time ago, and probably have become 'good species', although we know of no attempt to cross-breed them.

So speciation is a gradual process of divergence, until not only the form and behaviour differ, but the very structure of the genes has changed too far to allow successful recombination. The colour differences between the two giraffes and between the two zebras are probably ones which have come about by pure chance – a judgement admittedly arrived at by examination through human eyes, to which both types of blotches and both stripes are equally effective.

Herbivores: Pursuit of food

# Reaching a Niche

The **giraffe** is a product of evolution which has solved its problems by elongation. The long neck opens up a largely unexploited grassland food niche to the giraffe, putting highly nutritious leaves within reach. A food source twenty feet above the ground is virtually unavailable to most other mammalian herbivores. 'Browse lines' on the underside of a woodland canopy, or shrubs shaped like hour-glasses, are evidence of the hedge-clipping activities of giraffes, who browse on all the suitable trees in an area, taking just a few small sprigs from the surface of each. Like pruned hedges the trees respond with a thickening of the surface which in turn provides a greater leaf table area for the giraffe.

A neck of such unusual length demands that the heart of a giraffe be relatively large in order to pump blood ten feet uphill to the brain. When a giraffe drinks or browses from a ground-level shrub, the brain would burst from fluid pressure if it were not for special valves in the neck arteries. In turn, however, maintaining the elasticity of these valves and of the arterial walls, requires a diet rich in just these materials that are most plentiful in the vegetation at the very tops of trees, and which cannot be reached without a long neck.

The long tongue of the giraffe is just able to slip between the acacia's array of defensive thorns and pull the foliage into the mouth, lengthened and narrowed for the same reason. What thorns are ingested do not seem to bother the idly munching animal, partly because of a horny skin covering his palate and partly because they are coated with a special latex-like saliva.

The leverage and muscular masses of the long legs can kick with a force that puts giraffes near the bottom of the list of favourite lion food, and the giraffe's visual advantage in spotting danger before more conventionally structured prey animals makes him a valuable and respected member of the herbivore community. When a giraffe stops to look, others freeze and listen.

Herbivores: Pursuit of food

# Improbable Relatives

There are few mammal herbivores which could appear less alike than a 4 pound **rock hyrax** and a 4,000 pound **elephant**. Yet, physiologists can demonstrate a number of physical similarities and paleontologists can trace a common ancestry back to the antecedents of a forty million year old beast called moeritherium. These animals browsed and grazed their way through Eocene ('dawn of the recent') pastures and gave rise to progeny who solved different problems in very different ways.

The elephant reaches for food with his trunk, which is a remarkable evolutionary fusion of nose and upper lip. It effectively increases an elephant's upward and outward reach, but its main function is to enable the animal to transport food from the ground to its mouth, a distance of six or seven feet. Although elephants feed primarily on low grass and herbs, any vegetation from ground level to about 18 feet is fair game for the trunk. Even if the leaves of a tree are beyond that limit, an elephant can simply push the tree over to bring them to a more convenient level.

The hyrax almost matches the elephant's vertical reach by a remarkable agility for one so squat. One of the three East African species, *Heterohyrax*, is able to climb onto the flimsiest acacia branches and browse there almost as if grazing. The pads and toes of the feet, not unlike those of the elephant in internal structure, are covered with skin that actually sweats, which increases traction and surefootedness.

The hyrax achieves thermoregulation with a behavioural repertoire to augment its curiously inefficient internal thermostat. When the outside temperature drops, they seek the sun or flatten themselves against a rock which has been warmed; at night they sleep in piles in rock crevasses, insulating one another and decreasing the effective surface area of hyrax exposed to the cool evening air. They have a remarkable interwoven network of veins and arteries in their legs which ensures that the body heat in the outward-bound arterial blood is transferred to the cooler venous blood returning from the feet.

Paradoxically, the very size of the elephant means it needs less to stay alive than you would expect. A rotund animal, with a relatively sluggish metabolism and a large mass compared to its surface, requires less food per pound of body than a smaller animal. Small animals need a higher metabolism – a faster burning body furnace – to maintain themselves, partly because relatively more energy is lost in the form of heat from their surface areas.

In the heat of the tropics, there must be, of course, some compromise between making the most of the energy eaten, and dissipating enough to avoid succumbing to heat exhaustion. The elephant's superlative proportions serve this compromise. Physiologically, elephants avoid overheating by a slow metabolism, by the insulation of the one inch thick skin on the back, and by the great mass of the body which takes a long time to heat up to a critical level. Behaviourally, elephants cool themselves by seeking shade at midday, by feeding in swamps with their feet in the water; and by flapping their ears. The elephant's two-square-yard ears are heat-exchange devices. They are almost constantly in motion, and as warm blood from the deep body flows along the veins of the ear, it quickly loses heat through the very thin skin to the moving air outside, and returns to the rest of the body several degrees cooler.

The fact, then, that elephants are thermally efficient animals makes them efficient feeders, which in turn makes them successful. One large 6,000 pound elephant will actually consume less vegetation than the same biomass of smaller species – a large herd of impala, for example, or an entire population of hyraxes.

Herbivores: Pursuit of food

# Reaching for Food

Reaching for food is one of the most important things an animal does. Fundamentally it involves a sequence of movements that puts the appropriate part in the right place. As we have seen, this may entail the use of a highly specialised part, like the giraffe's neck, the elephant's trunk or the aardvark's sticky ant-catching tongue.

But reaching may also require a certain type of behaviour on the part of the animal that increases the effectiveness of its existing equipment – the **gerenuk** goes up on its hind legs to augment its elongated neck (top right); the **warthog** goes down on its knees to compensate for its short neck (bottom right). Or again, reaching may demand a generalised piece of physical equipment capable of coping with a variety of special occasions. The elephant's trunk is an amazing tool considering it is basically a nose, but the primate's hand is a thing of wonder. Perfect for omnivores like **baboons** or man, it is strong enough to kill and tear, dexterous enough to peel and pick. Not only is it used to prepare and transport food, but to manipulate, heft, examine and experiment with parts of the environment. A perfect appendage for a brain becoming curious enough to be called intelligent.

## Herbivores: Social organisation

# The Solitary

It has been said that an individual organism is the gene's way of making more genes. A society is one aspect of this, for the ultimate function of sociability is to increase the chances of survival of the individuals who participate. The possible types of social organisation form a continuous series of orderly collections of individuals, from the nearly solitary rhinos to the highly gregarious wildebeeste. It takes a minimum of two animals to be social, either on a permanent basis like dik-diks, who pair for life, or temporarily like a male and female rhinoceros who come together only when the female is in oestrous. Either way, it is a positive attraction of the animals to one another which maintains the social group. A group of vultures on a dead wildebeest is not strictly speaking a social unit, since the 'group' is formed by individuals attracted not to each other, but to an external stimulus.

Social communication is vital in maintaining the integrity of the group. It involves using a set of learned or innately understood signals – visual, auditory, olfactory or tactile. The language is predominantly species-specific, but some messages cut across species boundaries – a lion understands an elephant's threat, a rhinoceros reacts to a tick-bird's alarm call. Within a species, the messages are unambiguous (to the animals) and hold the group together with such meanings as 'I am here, where are you?', 'keep back', 'come here', 'stay still', 'follow me'. . . .

The nature of a species' food resource determines, in general, the form of social organisation it has. An abundant food supply means that the animal can afford to be more social and live in groups. A rare and sparsely distributed food supply would soon be depleted by a group. Thus at the least-social end of the social continuum stands the **black rhinoceros**. It eats herbs and browses on shrubs, food resources which are scarce compared to grass. Its large body size on the one hand means that many rhinoceroses could not get enough to eat in the same small area. On the other hand, its size and unhesitating willingness to use the fused elongated hairs which form its horns, mean that it does not miss the safety-in-numbers advantage which social groups enjoy. The fact that it is a lumbering bad-tempered recluse makes its only socialising, that of courtship and mating, seem rather ridiculous.

The male follows the oestrous female doggedly around his territory, he responding to her scent, she essentially communicating by her actions that he should keep his distance. Bouts of nudging, play-fighting, urinating and thrashing the ground with his horn eventually communicate to the female that his temper is suspended and it is safe for her to get close enough to mate. Copulation itself is unique in that it can last for half-an-hour, a herbivore record which may explain the pharmaceutical qualities attributed to rhinoceros horn. After mating, she may accompany him for a couple of months, or she may not. Thus ends the rhinoceros social season.

Herbivores: Social organisation

# The Family

**Burchell's zebras** have two levels of sociability: we very often see them in large herds, particularly in the wet season, migrating perhaps with the wildebeeste onto a locally abundant food source. But if we watch long enough, or especially if a predator threatens, we see the herd fragment into the basic social units – small mobile family groups of up to a dozen, comprising a male, his females and their young.

This response to predators indicates a key function of the family form of social organisation, namely protection of the breeding stock. A male zebra is a formidable fighter who will lead his family in seeing off a hyaena, and has been known to keep lions distracted while his family escapes. Such encounters are often fatal for the zebra, but as a result, some of his offspring escape and the sacrifice is worthwhile in terms of evolution. In the case of the zebra, it is not so much the constraints of the next lower trophic level, the food supply, which dictates the form of his social organisation, but rather more the limits imposed by the next higher level, the predators.

The perennial mystery of the function of the zebra's stripes might be explained in part as a by-product of its gregariousness. Obviously, we are tempted to invoke protective colouration, but the mode of deception is not clear. Certainly in dim light, from a distance, the form of the zebra dissolves in a blur where the stripes meet the body contour. Since predators like lions and hyaenas hunt mainly at night and detect their prey mainly by vision, perhaps the lost contours give a slight advantage to the zebra at a distance in dim light.

The stripes might confer a further advantage which depends on a group of running, mingling, boldly-striped beasts presenting a confusing image of a mass of vertical lines to the advancing predator – difficult to get a visual fix on one discrete form for the final leap. In short, zebras in a herd may form their own background against which each individual is camouflaged. Unfortunately, such hypotheses are almost impossible to test, and intelligent speculation will have to do for a while.

**Herbivores: Social organisation**

# The Harem

Every adult male **impala** aspires to hold a territory, to defend his square-mile of land from rival males, and to use the attractions of his resources to lure and keep female herds on his plot.

Within every **impala** population there are three types of social groups: herds of females and young who move from territory to territory; territorial males either accompanied by one of the mobile harems or temporarily alone; and herds of bachelor males.

The bachelors maintain a strict hierarchy within their herds. Young adults give way to the older animals, and the relative positions of all individuals is established by sparring. But the system is in a constant state of flux as new animals join the herd, and the dominant adults leave to establish or take over existing territories. While he is an unattached bachelor, the male has little to distract him from keeping in top condition. He may spend eight hours of every twelve just eating and idling. Once he achieves the status of territorial male however, the situation changes drastically.

Territorial boundaries are marked with the secretion of a gland of the forehead which the male rubs on border bushes and trees. The male himself serves as a territory marker by posturing and vocalising at the boundary. He will defend his area against all male intruders, but territorial disputes are rarely violent. They more often involve display rather than actual physical combat, and the weaker animal will invariably retreat. This mechanism ensures that only the strongest males are able to father young. Equally important, it also means that the females are served by a variety of males – thus maintaining a reservoir of genetic variability in the population.

The impalas' speciality among the African antelopes is an ability to use a wide range of food plants – this is their particular strategy for dealing with the seasonal changes that dominate the cycle of their year. During the wet season impalas are thinly dispersed

throughout all suitable areas, feeding on a generalist's range of green vegetation – grasses, herbs, bushes, trees and fruits. They need shade too, and so avoid treeless country.

As the dry season advances impalas retreat to the woodlands and river valleys where the vegetation is still green, a retreat that is accompanied by a progressive breakdown in the territorial behaviour of the males.

In the wet season the population density of impalas in their habitat is unlikely to be more than 50 to the square mile; but the animals will be found in compact groups, the individuals feeding within a few yards of one another. Dry season ranges are much smaller in area, and densities will be as high as 250 impala per square mile. The woodlands are able to support this greater concentration because they provide a diet with a relatively higher protein content. However, since this diet is predominantly browse rather than ground vegetation, individual animals will be more widely dispersed while feeding than is the case on the more open wooded grasslands. Thus individuals are most tightly packed in groups when the population is most dispersed (wet season) and most spread out when the population is most concentrated (dry season).

A territory can have a variety of functions, depending on the species. For the impala, there seems to be a premium on keeping the browsing load spread evenly over the habitat. If there were no mechanism to keep hungry female herds scattered, the best areas would be continually over-crowded with impalas and over-eaten. Vegetation needs a chance to recover from concentrated herbivore predation. A territorial male is essentially preserving a part of the population food reserve – small enough to effectively defend and large enough to attract and nourish the female herds which wander into his sphere of influence and which will ensure continuation of his line.

# The Matriarchy

During the sixty-odd years of an elephant's lifetime, it will wander over a range of 10–20,000 square miles and experience perhaps five major droughts. This means it has a lot to learn: where there is water; when different food plants come into season; where, in the driest times, some food and water can always be found; and how to avoid its only predator – hunting man. It would be inefficient, certainly dangerous and most likely impossible for a young elephant to learn all the tricks of survival by exploring its habitat alone. Better to draw on the knowledge of older elephants: to be taught, in other words.

But then again, the secrets of a habitat which must be coped with for sixty seasons cannot be learned in one, or even five; thus **elephants** have the longest childhood of any animal, except man, and stay with the family until puberty, perhaps for fifteen years. The closest tie a young elephant has is with its mother – she is its earliest teacher. For the first year it will barely stray a trunk's length away. It tastes the food in her mouth and suckles for as long as six years. Not surprisingly then, the elephant has evolved a matriarchal society. The basic social entity is the family unit led by one mature cow, the matriarch, who is either the mother, elder sister or cousin of everyone else in the group.

Female calves will stay with the family group long after puberty. Young bull calves, however, begin to stray at that age, primarily because the cows become intolerant, even to the point of tusking them out of the group. This ousting has the function of preventing future interbreeding and also works in favour of the family unit, since there will be one less mouth to feed in the dry season. And it benefits the young bull himself. He may be sexually mature at fifteen, but he will not become 'socially mature' until about twenty five, when size, experience and strength allow him to compete for cows in heat. If he stayed with the family group, he would be subjected for almost a decade to the competition of courting bulls, mostly larger than himself. He would continually be giving way to the larger animals, and have little opportunity to spar and test his strength with his contemporaries in bull groups.

Bulls do not form permanent groups, but rather loose assemblies of from two to fifty individuals with no long-term social ties. Like all animal herds, they have a primary function of safety in numbers; but beyond this they allow the young animals to meet the other males in a population, and to judge when they are ready for serious contest for the females.

It appears that bulls are also the population pioneers. They are more frequently found on the periphery of the population's range and invariably precede the female herds into new pastures. Again we can see an evolutionary design in this system. On the one hand they are physically more able to explore new regions, since they are larger, stronger, and can go longer without water than the young animals of the female groups. And, in terms of the population's capacity to perpetuate itself, they are more expendable. The loss of a bull is little more than the loss of one elephant. The loss of a cow, however, is not only the loss of a teacher or even a group leader, it is also the potential loss of ten elephants – for optimally, a cow will produce a calf every four years between her fifteenth and fifty-fifth year.

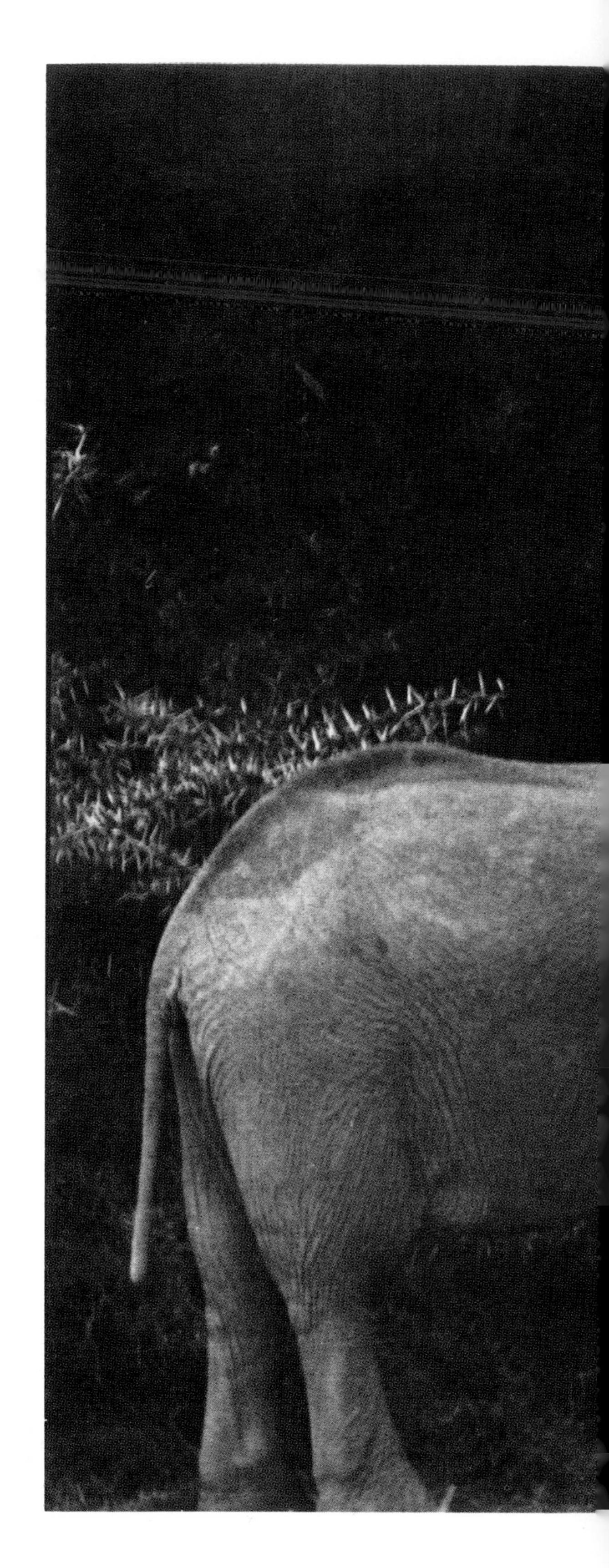

## Herbivores: Social organisation

# Herds and Troops

The first reactions of a herd of **buffaloes** or a troop of baboons to an intruder are an indication of the key function of their social organisation. The adults of the mixed 'families' in these groups look and listen and smell the disturbance. The baboons sit up and face the intruder or climb a tree for a better look. The buffaloes stop grazing and point eyes, ears and noses at the intruder. Some may even advance a few paces with heads raised, giving the impression of myopic curiosity. Young animals move to the rear.

A collection of sense organs detects and examines the discordant stimulus. A consensus of experienced adults appraises the situation and a collective decision is made whether the best strategy is to attack, flee or resume feeding. Subordinate animals may

take the cue from the dominant ones, which is not surprising since the animals near the top of the hierarchy are usually the oldest, who have necessarily passed a long series of tests and choices that the environment presents to them. There is early warning, as well as safety in numbers.

Many types of food resource occur in patches so that feeding automatically produces herding in the first instance. Moreover the relative availability of food may be increased if more sense organs are applied to detecting it. A baboon that discovers a bush with ripe fruit will eat the find quickly if there is not much of it. But if the bush is laden, his actions and excitement will attract his fellows to share the meal. Or the advantage may be a more complex one: the trampling, grazing and excreting of many buffaloes in valley grassland promotes the growth of young, nutritious shoots to the obvious advantage of the whole herd.

The mixture of families, the combination of both sexes and all ages, the physical proximity of members, the co-ordinated activities and resource sharing in herds and troops are rather like human societies. Even if all the individuals do not know one another, which in fact they probably do, status is instantly recognised by subtle signals, akin to those that humans respond to among their own species, but often are not aware of. Body posture, tilt of head, vocalisations, direction of gaze, perhaps even body odour are used to signal intent or rank in all species. The result is a relatively tranquil order in social groups, which ensures less energy is used in bickering and more in surviving.

# The Underground Society

While we watch a herd of buffaloes grazing we are usually indifferent to the activities of insects. Yet insects – like locusts which eat live plants and termites which consume dead ones – probably transport more materials than the spectacular large mammals of Africa. **Termites** in particular return tons of nutrients to the soil by literally carrying dead wood and grass below ground. This would otherwise decay far more slowly in the dry grasslands.

Dotted at regular intervals over most of the grasslands are termite mounds – at once dumps of earth excavated from the subterranean tunnels as well as cunningly designed flues for allowing heat to escape. It is a marvel of instinctive behaviour that the mounds are all so similar in appearance, for they are built by millions of workers, none of whom ever see the finished product as we do.

Termites are not, as we might think, related to other social insects such as bees, wasps and ants. This mis-named 'white ant' is not an ant at all: it belongs in a group of its own, more closely related to cockroaches than to ants. They have no forms which develop from unfertilised eggs, as worker bees do. Instead of starting life as a larva, the young termite hatches directly from the egg into a nymph (lower right), an immature copy of the adult.

Some termites are actually able to eat wood with the help of minute single-celled organisms in their stomachs. The so-called 'gut flora' produce chemicals which break down the tough plant fibres and render them digestible for the termites. Other termites like *Macrotermes bellicosus* (the 'big warlike termite') have no such helpers. Their answer is to become agrarians and cultivate their own food.

Underground, in the heart of the lightless chambers, *Macrotermes* shapes the earth into convoluted hanging gardens seeded with the spores of a particular fungus species and fertilized with half-digested regurgitated wood pulp. The little mushroom-like growths which flourish in the dark (right, second from bottom) are the real food of *Macrotermes*.

In the rains when the soil is soft and moist, a caste of males and females becomes sexually mature and sprouts wings. One evening, after a heavy shower, workers open tunnels to the surface and the winged emissaries fly up towards the failing light (top right). In the darkness they drop to earth again and having shed their wings perfunctorily, males and females form pairs and run off in tandem to seek a suitable site for the beginnings of a new nest. Those two creatures are all that is necessary to form a new colony; the female even has in her gut a minute piece of fungus to be regurgitated and become the nucleus of their fungus gardens.

Of the untold thousands that flew from the one nest only a minute fraction will survive the depredations of numerous predators and live to establish another colony. And the thousands that die – are they wasted? The answer must be, no. Termites are extremely tasty fare to predators, and on their nuptial flight they are at considerable risk as they flutter in silhouette against the evening sky. But there are so many of them; probably more than enough to fill the local predator population who cannot cope with these seasonal excesses, since their numbers are controlled by the average amount of prey available all year round. The enormous sacrifice ensures that some termites survive; and in any event, all the material will be recycled via bird droppings or genet dung and will be used again one day by termites. Thus for the ecosystem the nuptial flight of the termite ends in a rain of nutrients.

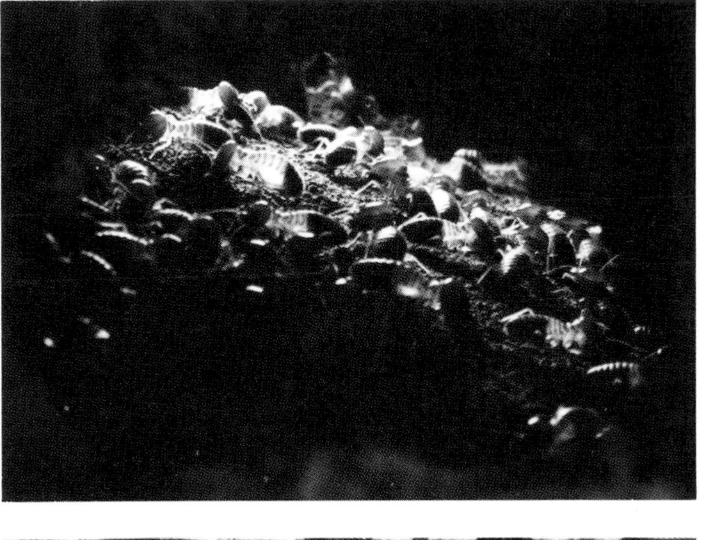

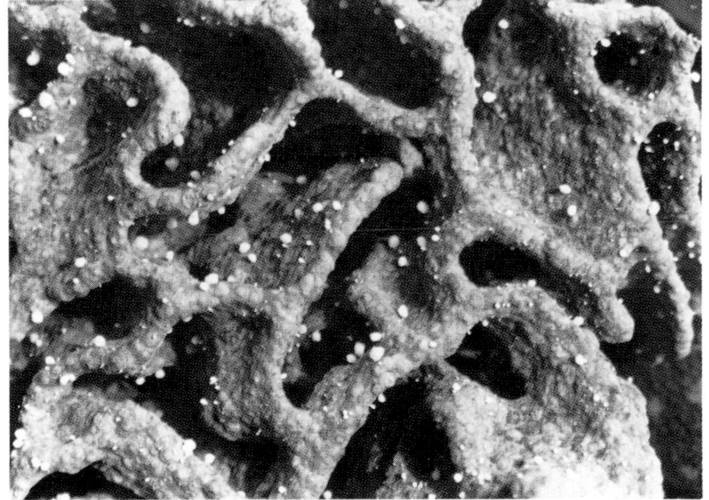

Herbivores: Reproduction

# The Royal Family

A termitary operates on a caste system, a strict division of labour among builders, soldiers, foragers, cleaners and nurses. There may be several million individuals in one colony, but there is no thought process or reasoning among them: each works or fights instinctively, according to the genetic information programmed into it.

The remarkable order and efficiency resulting from this fact has led to the popular conception of the termite colony as a 'being' in itself – a sort of 'super organism', in which the individual termites are analogous to cells and the duties of the castes to physical functions: the foragers are the blood stream carrying nutrients through the organism; the cleaners are its excretory organs; the soldiers, its claws and fangs; the king and queen, its sex organs. It is a compelling metaphor because it *describes* a termite colony so well; but it does little else. We would like to know *how* that colony works.

Three feet or more below ground, secure within strongly sculptured walls and fiercely protected by thousands of aggressive soldiers, lies the **queen termite**. The very same creature that flew delicately into the evening light has become a grotesque four inch egg factory capable of the astonishing production of 30,000 eggs a day for twenty years. The king is always beside her to inseminate each crop of eggs, though we do not know if he is necessarily the same individual with whom she paired after the nuptial flight.

Attendants constantly stroke and cleanse her abdomen; as her body ripples with the convulsions of egg laying, a continuous stream of workers passes through the cell, bringing food and carrying eggs away to the nursery chambers.

In the nurseries, eggs hatch into nymphs, which, remarkably, are capable of becoming members of any caste. It has been shown experimentally that precisely which caste an individual grows into is somehow determined by the needs of the colony itself. If a colony is in an early stage of construction, most nymphs will become workers; if it has a poor defensive structure, more will become soldiers. Many simply remain nymphs until the needs of a particular caste become apparent. Apparent? How does an instinctive, genetically programmed, totally 'unthinking' colony of insects perceive its needs? We simply do not know. But termitologists contending with the enormous problems of observing a functioning, undisturbed colony are enquiring into the mutual feeding behaviour of termites as one of many lines of research. Perhaps there is a *taste* of social state, instigated through hormonal messages transferred in the food – the original word of mouth.

# Male Rivalry

Breeding is the class of behaviour concerned with maintaining a species' integrity. No matter how much there is to eat, a great deal of the energy consumed must go into the activities designed to maintain the species or the food is wasted. On the other hand, no matter how prolific a species may be, the proliferation will snuff itself out if there is not enough energy to feed it.

A first step in the transmitting of genes is a decision about who will do the transmitting. If all mature animals were to produce young, the population's progeny would contain a considerable portion of substandard material, simply by the laws of chance. A genetic line is constantly at risk: it can lose the struggle against an unfriendly environment at any stage in an animal's life cycle. Some stages are more vulnerable than others – when new-born, when very old, under conditions of stress in the dry season. The risks can be minimised and the energy going through a population maximised if the parent's contribution to the next generation is of the highest quality. Hence, male rivalry.

Males, particularly male ungulates (hooved animals) like **Grant's gazelles** (top) or **wildebeeste,** invest a good deal of effort in deciding who will mate. They fight to establish a hierarchy, they fight to maintain a territory, they fight over females. The decision process ensures that only the strongest, most tenacious, quickest and most cunning will pass on their traits. The screening process of aggressive male contests increases the fitness of the population as a whole by letting only the fittest genes through.

Of course, a male ungulate sporting sharp horns, driven by his genetic vigour, fired with a seasonal extra dose of male hormones, is a very dangerous beast. Yet fights are rarely fatal. Clearly, nothing could be more disadvantageous if practised on a large scale. Thus there have evolved conventions, rituals, almost sporting tests of strength which are just that and not frenzied battles to the death. If contests were fatal, we could imagine a population of wildebeest whose most blood-thirsty males, capable of killing lions, would soon dwindle through self destruction. Just as legs are neither too long nor too short, horns neither too unwieldy nor too small, so most behaviours have a safety valve, a built-in stop which inhibits over-reaction. Gentlemanly behaviour is adaptive, murder is not.

# Courtship Ritual

A male herbivore that has survived till sexual maturity and has established himself with his peers near the top of the hierarchy, has spent most of his adult life keeping his distance from other creatures – fleeing from predators, separating himself from rival males. Females share his aversion to closeness. If he is to fulfil his task in life successfully – to produce children – his whole attitude towards physical contact has to change, particularly with respect to females. Sex hormones, whose production may be triggered by environmental factors such as change in day length, temperature, food availability, or the presence of a mature female, take care of the attitude of the male well enough. But then the problem is to convince the female that this excitable creature confronting her is not dangerous. Not only must she not run away, but she must stay still, and at least for one vital instant be available for the act of mating. Thus we have courtship, an often elaborate combination of attraction, appeasement and reassurance behaviour.

The male **Jackson's widow bird** is a gallant, if slightly mad, suitor. He sets up a territory among his fellows in open grassland, and on it performs a circular leaping dance with such vigour that he eventually wears a trampled ring in the grass. If this bizarre display can attract the attention of humans, it can certainly arouse the interest of females evolved to respond to it. Flocks of females fly over the display grounds as if appraising the performances of the cavorting

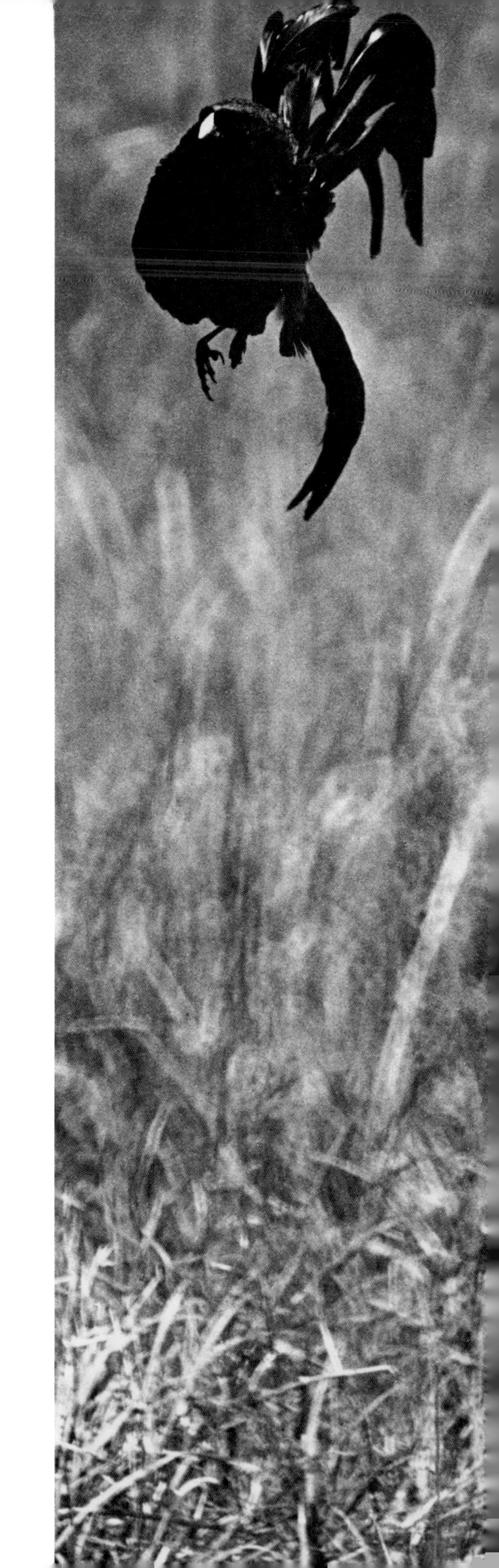

males, a dozen of whom may be jumping about below. When a curious female makes her choice and lands in a ring, a brief bout of hide-and-seek around the central grass tuft follows. The tuft allows her to shield herself from the awful splendour of her suitor; but the ring makes the outcome inevitable.

For most of the year there is little to tell between the male and female Jackson's widow birds. The male is slightly larger and a shade darker, but by and large both are insignificant: hardly more remarkable than the suburban sparrow. But in the breeding season the male looks like an entirely different creature. His plumage is a rich shining black with contrasting olive-brown shoulders, his tail is thick, long and curved. In overall effect he bears about as much resemblance to his other self as a rook does to a thrush.

The dowdy year-round plumage is good camouflage and, as such, especially important to the female while she is carrying eggs. But to ensure she is fertilised in the first place, the male must abandon the security of his camouflage at the start of the breeding season and attract females. There is a risk of course, that he will attract predators too, so he wears the striking breeding plumage for no longer than necessary.

This dimorphism (two forms) and dichromism (two colours) is a tactic that many birds have evolved. It is the ultimate species compromise between advertisement and concealment.

**Herbivores: Reproduction**

# A Time to Mate

Animals come into season because the season gets into them. Annual changes in things like wetness, dryness, day-length, grass texture and content, cause the animals to begin breeding activities. Surprisingly, we do not yet know for certain which of the many possible stimuli actually trigger breeding of the large African mammals – a statement, in fact, which is true for most species. Wildebeeste and gazelles court and mate in the dry season. Elephants increase their sexual activity during the rains. Giraffes mate all year round.

Males usually confirm by scent what the hormones of their activated sex glands tell them, or what the females convey to them through their behaviour. The male smells the female's genitals or tastes her urine for a chemical message of readiness.

Almost all ungulate males react like the grant's gazelle to the smell of a ready female with a head-forward, lips-curling posture called 'flehmen' (top left). We are not sure what the posture means. It is not likely to be a signal to the female: she usually does not even notice. Possibly other males may respond by avoiding the couple, although flehmen will occur with no other males about.

Although the timing is critical for the actual mating, the act on which humans place such a high premium is a cursory event among animals. It is but one of the several steps in the process of gene exchange, and may last only seconds between gazelles and less than a minute between elephants. The only large beast who lingers over the event is the rhino, and the one who most conspicuously loses himself in transports of orgasmic delight is, curiously, the ostrich.

The point of seasonal breeding is usually to deliver young into a world which offers an adequate food supply. Animals of different gestation periods must clearly start their breeding activities at different times if they are to all take advantage of the same rainy season. The wildebeest has an additional reason. The births of the calves are closely synchronized so that in February, there may be over half a million new born wildebeeste on the Serengeti plains. The calves are extremely vulnerable to hyaenas, wild-dogs and the cats. But the synchronized calving has the effect of glutting the prey market: the same strategy the termites use on their nuptial flight. The predators spend much of the period lolling around with full stomachs, literally unable to eat more. Consequently, they eat less of the wildebeest calf crop than they would if taking a constant toll the year round.

Herbivores: Reproduction

# A Deliquescent Nest

The change of seasons, from dry to wet, happens almost overnight. The half a year of rainless weather finishes with a crescendo of atmospheric tension and stifling temperatures. And then it rains – where there were dust and sand yesterday, today there are pools and rivers. Some of the trees and large mammals have already begun the internal reproductive changes which synchronize their cycles to those of the land and sky. Smaller organisms respond immediately to the rain. Grass begins to grow again that very day, insects hatch, and from out of the parched ground of the bushed grassland in Tsavo National Park creeps, of all things, a frog.

It is never certain that water will remain very long, particularly at the beginning of the rains. So the frogs hurry to the edge of a waterhole within a few hours of its forming. It is not an event we are likely to see, since the rains usually start at night, probably because it takes a day's heat to build up the clouds to a suitable state of instability. A female frog is joined by one or more of the smaller males, and there begins a frenzied mating scene in which the female produces eggs and a mucus which is worked into a foamy mass by the paddling hind legs of both animals.

At sunrise we find the frogs sitting atop a soufflé of mucus, eggs and sperm. Even if the water hole dries up before the next shower, the eggs can hatch in the guaranteed moisture of the foam. At the next rain storm, the deliquescent nest together with the tadpoles is washed away into the pond. By this time, in a normal year, the rains, will have begun in earnest, and the pond should last the few weeks necessary for the tadpoles to grow into adult frogs.

How or where **Peter's tree frog** survives the dry season is not known, for it must aestivate, which is a hot weather version of hibernation. It is an amphibian, after all, which must in theory have moist skin to stay alive: much of a frog's breathing is done through the skin. But how does it prevent total evaporation of its body fluids?

Following heavy rain, other dry-country relatives of the foam frog have been observed to rear up on the still-wet earth and literally disappear into it as one watches. They burrow backwards, as far as the earth is wet, which may be four feet down. There they make themselves a mucus lined chamber, which dries into a seal and protects them during their dry season 'sleep'.

# Husbandry at the Nest

There is nothing very special about an ostrich nest – a perfunctory scrape in the ground, made by a male in his territory – similar to that of most other ground-nesting species. The all-white eggs may indeed be dangerously conspicuous, but the camouflaged, mottled brown female covers them during the day, the black male at night, and both 200 pound animals are formidable in defence. No other bird can claim the distinction of having kicked a lion to death. In any event, most potential predators, lions included, are unable to break an ostrich egg.

The **ostrich** seems to carry egg production to excess. Nests with over 100 eggs have been recorded. Since the sitting bird can only cover and hatch about two dozen of these (about the same as a successful goose), we may understandably wonder why so many eggs are necessary. Part of the answer is that ostriches are polygamous: the male invites, courts, mates with and accepts the eggs of more than one female. The females, for their part, lay eggs in more than one nest. Also, ostrich eggs are relatively inexpensive to produce in terms of energy and materials: a song-bird egg is about one-tenth of the adult weight, an ostrich egg about one-sixtieth. But we are skirting matters of mechanism of the large clutch size – the 'how' of mechanism rather than the 'why' of function.

We might approach the function question by stepping back and looking at the ecology of the bird. Ostriches are selective feeders, who, like elands, rather fussily pluck small, green shoots of wild flowers out of the tangle of a grassland. A scarce primary resource means, as in the case of rhinos and elands, a low population density. In addition, a well-defined dominance hierarchy among the females, results in the same old experienced hen sitting on a particular male's nest every breeding season. Thus, a particular pair would produce offspring with the same genetic make-up year after year, if there were not a special mechanism in their breeding strategy which allowed other females, as it were, to contribute their genes to the pool. Through the egg lottery at the nest, the offspring of a breeding season have a high genetic variability, a healthy thing for any population in an environment in which climate and primary production can fluctuate as much as they do in the tropical grasslands. Moreover, if one nest is destroyed by, say, a flooding rain, at least some eggs of any one bird will have a chance of surviving if they are laid in several nests.

But having said all this, it must be admitted we are still in the speculative stages. It will take further careful research involving marking both eggs and birds to solve the mystery at the ostrich nest.

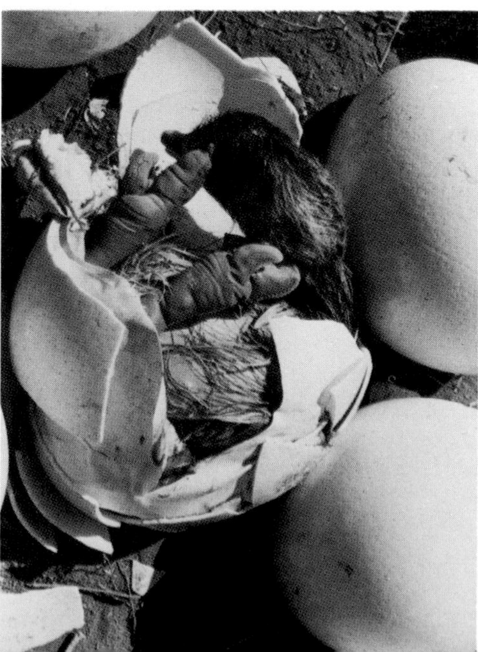

Herbivores: Reproduction

# The Economy of Numbers

We would expect a herbivorous bird like the **ostrich**, which hatches fully-feathered, mobile, ready-to-feed chicks to produce a relatively high number of them. Primary production is abundant and immediately available to the hatchlings. The conservative one- or two-chick economy of, say, a bird of prey, who must defend and keep warm the young, as well as travel miles to collect enough food for them, is not a constraint in the ostrich's reproductive strategy.

But whereas ostrich eggs are easy to produce, the chicks are difficult to defend. Their small size makes them fair game to a wide spectrum of predators – from eagles to lions. Certainly adult ostriches are effective defenders of their brood, but it is a formidable task to keep a group of 20-odd chicks together when under attack.

Consequently, ostriches have evolved a unique solution. There is a fairly precise synchrony of hatching within one ostrich population's range, and most chicks follow their parents away from the nest in the same week. Thereafter, when two broods meet in their wanderings, they join together in a confused shuffle. The dominant pair of adults assert themselves over the other pair and end up leading both broods away. This continues until almost all of the broods are under the control of one pair of adults. There may be up to one hundred chicks in the group.

One can hypothesize a number of advantages as well as some disadvantages to such a system. Imagine a predator wandering about looking for something to eat. If there are ten ostrich broods in his range, there are ten possible targets he can encounter. If the broods join into one or two large broods, the effective target area is less and so is the chance of a 'hit'. On the other hand, a large group is possibly more conspicuous and slightly easier to detect. Then again, if the predator usually gets one chick in an attack, before the adult birds get the rest safely away, then *each* chick in a large brood has a greater chance of survival than *each* chick in a small brood. One could go on with such a list, but clearly, the advantages must in the long run outweigh the disadvantages.

The lead male who finds himself burdened with all of the chicks, some of them sired by males against whom he has just spent a couple of months defending his territorial boundaries, is not just being a good neighbour. It is difficult to believe he would incubate and defend all of these chicks unless most of them carried some of his genes. We may, one day, no longer talk of an ostrich population nor even a colony, but of a super-family, spread over hundreds of square kilometers.

**Herbivores: Avoiding predators**

# Symbiosis

We usually think of 'co-operation' as being a relationship which involves a more or less conscious sharing of effort to produce a single mutually beneficial effect. Symbiosis is a form of 'pre-conscious' co-operation in a partnership which benefits two organisms, each in different ways. For example, an **eland** and an **oxpecker** (right) occasionally share the same space, the bird sitting on the antelope's back – a curious proximity at first glance. But the eland provides the bird with a pasture of skin parasites, and the oxpecker offers the eland a danger-detection system perhaps a shade more sensitive than its own. In this case the co-operation is more striking since it occurs between two creatures on different trophic levels.

Plants, of course, do not behave in the usual sense, simply because they cannot actively move in a time scale which we recognise as behaviour. But they do exhibit anti-predator, or rather, anti-herbivore adaptations. The thorns of the African Acacia trees are designed to discourage browsers and therefore decrease the 'predation' on vital photosynthetic material. Australian acacias in contrast, have no thorns, presumably because they evolved in the absence of browsing herbivores. Another example of how comparative observations can give us hunches about function.

The **whistling thorn** acacia tree has bulbous galls at the base of its spines. With a few doors and a little hollowing out, they provide excellent homes for a small but pugnacious species of ant (*Crematogaster*). The ant gets shelter and a thorn fence, the tree a second line of defence after its armament. Hungry herbivores get bitten as well as pricked if they try to eat *Acacia drepanolobium*.

The co-operation has proved so successful to both species that they have virtually evolved together: it is rare to find whistling thorns without the ants, and the ants never occur without the trees. Ants can, of course, live just about anywhere, but how much more beneficial to have a living home which can evolve structural improvements to accommodate the tenants. The fluting tones produced by the wind blowing across the ants holes in the galls are delightful to the human listener, but are, of course, incidental to the symbiosis.

Essentially, a symbiotic relationship involves two species doing more or less what they would do in any event, but tolerating the close presence of one another whilst they do it. The toleration, then, which costs virtually nothing in terms of energy expenditure, increases the chances of survival of both.

**Herbivores: Avoiding predators**

# First Lines of Defence

The third major concern of all organisms – after eating and reproducing – is avoiding getting eaten. Of course, both the organism which escapes predation and the one which gets enough to eat have a better chance of breeding successfully. The energy necessary for behaving or for building specialized body parts is most frequently expended in the processes of feeding, anti-predation and reproduction.

The head of a herbivore is a conspicuous indication of his way of life. A main feature is the danger detection apparatus. The herbivore early-warning system consists of a large nose linked internally to a complex honey-comb of scent-sensitive membrane, capable of detecting molecules in the air and identifying their source with a sensitivity unknown to us. There are the large, well developed ears, swivelling sound baffles. And there are the eyes.

Ungulates do not need their eyes to locate food. They can smell it well enough, and to close accurately the gap between food and mouth they are aided by the touch-sensitive whiskers on their lips. But the eyes are essential for detecting danger, and to give the best field of view they are usually located on the side of the head. The **dik-dik** (left) or the **african hare** (lower right) can see what is behind almost as well as what is in front.

After early warning, the next line of defence against predation is simple inaccessibility. Camouflage colouration effectively makes the prey visually inaccessible to the predator. We have already suggested that the zebra's colouration, combined with gregariousness, has the effect of reducing the visual availability of any one zebra in a fleeing group by confusing the predator's target image. 'Flash' colouration is another way of confounding the visual predator. Some ground-coloured animals like the **bushbuck** (upper right) or the hare, have conspicuously white tails which are flashed intermittently at the pursuing cat. This gives the predator momentarily a clear signal on which he can fix, but one which vanishes in an instant. In the dark, we are similarly disorientated after a light is flashed in our eyes.

The last and perhaps the most obvious form of inaccessibility is simply hiding, going to earth. This strategy works best if the prey is smaller than the predator, who simply does not fit into the suitable retreat. It is not only to get at subterranean food such as roots that voles and mole-rats live in holes smaller than the various species of African cats. Nor is it simply to get out of the cold dry season night air that hyraxes pile into rock crevices narrower than the average leopard.

a
e
b
d
c
g
h

j

# Speed for Flight

If the herbivore's eyes, ears and nose do not sense the predator until it is too late to avoid detection, or if the predator sees through the protective colouration, or if he breaches the retreat, there is only one final recourse – to run. Energy used in flight is costly, but clearly well spent.

The fleeing prey and pursuing predator present what at first glance might be a Darwinist's paradox. It is obviously to the wildebeest's advantage to run a bit faster than the hyaena. Natural selection should favour a faster wildebeest. On the other hand the hyaena survival is enhanced if he can run down the wildebeest. So selection should also favour the faster hyaena. We might therefore expect an evolutionary race in which both prey and predator were producing faster and faster individuals. By now the contest should have reached a point where the landscape is ablur with supersonic prey and predators hard on their heels. Of course this fantasy ignores the multiplicity of selective pressures which result in any one final form or behaviour. Structural engineering problems – such as length of bone and stress in joints; physiological limits set by overheating and supplying blood to outsized muscle masses; nutritional problems and loss of agility associated with a large body – all act as evolutionary governers on animal speed. Compromise produces the situation in which today's cheetah and today's gazelle run as fast as they can, or ever could, in the current set of environmental circumstances.

a b c **Burchell's zebra**
d **Uganda kob**
e f **Thomson's gazelle**
g **African hare**
h i **kongoni**
j **eland**

**Herbivores: Avoiding predators**

# Weapons to Fight

The stag at bay and the wildebeest surrounded by snapping wild dogs are herbivores who have nearly run the gamut of their anti-predator adaptations. When retreat is thwarted, the last-ditch stand is the final recourse to save one's genes and materials for one's own trophic level. Poisonous snakes use their venom to threaten or incapacitate their predators as well as their prey. Herbivore horns have developed, primarily, for the purpose of self-defence so they must occasionally prove effective.

The impressive horns of an **oryx** may be, surprisingly, not only a defence mechanism but also an adaptation to conserve water. Most herbivores rely on running to escape from predators. A running oryx generates heat which has to be dissipated. That can use up precious body water reserves through evaporation. Such water loss can be fatal for an animal who lives near the climatic boundary of the grasslands, where the credit balance between rainfall and evaporation diminishes almost to zero. To get water where there are no water holes for most of the year, the oryx does much of its feeding at three o'clock in the morning when the withered vegetation has absorbed a little moisture from the cool night air. So, rather than trying to outdistance a lion, it may be cheaper in terms of ultimate survival to fight it out. Such an evolutionary decision requires defensive weapons, and an oryx is the only herbivore of its size which has been known to wound a lion fatally with its horns.

Although the function of the striking black-and-white facial colouration is not known, one would expect it to be a social signal – to conspecifics and would-be predators alike. The implied message of such a signal is obvious – 'a head sporting dangerous weapons is now turned towards you – beware!' Conspecifics avoid useless bloodshed, predators learn to retreat: effects worthy of an unambiguous signal.

Oryx horns or hyrax incisors would not be of much help without the will to use them. The stress and excitement of the chase increases the supply of adrenalin to the blood which results in untoward muscle performance and aggressive behaviour. The worm literally turns. The smaller jackal sees off a hyaena; the plover attacks the jackal; occasional success is clearly better than none.

Carnivores: Pursuit of food

# The Design of the Hunter

The front end of a carnivore is as conspicuous a clue to its way of life as the front end of a herbivore is to its particular nature. The form is dictated principally by the fact that carnivore food is usually mobile. Moving food must not only be seen and identified, but its speed and distance judged precisely to ensure an accurate lunge. The complex optical requirements for this job are partially met by eyes which face forward and fix the prey simultaneously. Anyone who has tried to touch the tips of two pencils together with one eye closed appreciates the advantages of binocular vision.

Once reached, the prey must be caught and held. Hence, predators have developed a range of hooks, claws and talons, and the strength to hang on and stay. Once held, the prey must be killed and somehow made manageable for the predator's gullet: both acts require teeth used in various ways. For killing, the lion or **lioness** usually smothers the wildebeest by clamping its muzzle shut, or like the cheetah, simply strangles the prey with a prolonged bite at the throat. The serval cat bites through the hare's spinal cord. Wild dogs and hyaenas tear their prey to pieces, so that killing and dismembering become the same process. Knife-edged molars slice up the meat, and in the case of the hyaena, grind up most of the bones too. The smaller the prey, the less reason for the nicety of a clean kill. One praying mantis can easily hold down a struggling fly, or one bird of prey a vole, whilst the meal is begun.

Carnivores' food is more concentrated than the herbivore plant fare. Carnivores consume more protein and energy, and less non-digestible material per time of feeding. Every pound of grass the wildebeest eats contains about an ounce of protein; every pound of wildebeest flesh which the lion ingests is nearly pure protein. One consequence of this is that herbivores spend most of their time feeding, while carnivores eat in spaced bouts and are sustained in the intervals by their high-quality diet. A wildebeest may spend eighteen hours a day feeding; a lion will spend the same amount of time sleeping.

Near the top of the terrestrial food chain our attention is drawn inevitably to the large cats of Africa – beautiful and terrifying, perfectly evolved hunting and killing machines. But, sharing the pinnacle of the pyramid of numbers are some twenty-seven species of mammalian predators, from lions and serval cats down to mongooses and shrews. There are many predator species quite simply because there are many herbivore species to serve as prey. There are large predators because there are large herbivores. Wildebeeste, zebras, buffaloes, hartebeeste and gazelles present the opportunity for lions, leopards and cheetahs to exist.

The nature of an animal's food supply dictates its whole life-style, limits its absolute abundance and may regulate its numbers over short periods. We have already seen that because of the constraints of the Second Law of Thermodynamics the weight of carnivores in a stable ecosystem will invariably be less than the weight of herbivores. Even within these absolute limits, when the herbivores grow scarce, either by dying or going away, some carnivores must necessarily starve. Although occasionally carnivores do keep a herbivore population in check, the more general case is that of the herbivore controlling the carnivore's numbers. A curious turnabout when the fate of the fearsome cat is dictated by the population dynamics of the timorous gazelle.

# The Solitary Hunter

**Cheetahs** walk like indolent queens, but when they run, their loose-jointed gait makes sense. They are designed to run; they hunt not by stealth or cunning like leopards, but by the chase. There are compromises which must be met for the distinction of being the fastest land animal. The head is small, seemingly out of proportion with the rest of the body. The reduction in weight presumably increases the speed and agility necessary to catch Thomson's gazelles, one of the easiest of the herbivores for the weak-jawed cheetah to kill. The specialization of speed severely limits the cheetah's range and confines it to open plains or sparsely wooded grasslands where there is a little cover for an initial stalk and enough room to run down small antelopes in sprints which last only a few minutes. Even though the heart beats slower than that of most animals of comparable size, the sort of physiological effort that the cheetah puts into the chase would kill it if of any longer duration. When the prey is knocked down and strangled after a hard chase, the exhausted cat may spend fifteen minutes regaining its breath.

They are usually solitary, or seen in small groups of two or three, perhaps a mother teaching her grown offspring how to hunt. The technique of tripping up a zig-zagging gazelle at sixty miles per hour has to be carefully observed and practised for months until mastered. It is probably the speciality of the cheetah's hunting technique which limits its numbers, even in a land abounding in gazelles and wildebeest fawns.

The cheetah is diurnal, partly because it is simply not possible to run very fast in the dark and partly to avoid competition with nocturnal lions, hyaenas and leopards. The spotted camouflaged coat probably serves to conceal the cat from larger predators as much as from its prey.

Carnivores: Pursuit of food

# Competition Between Species

It is not unusual to see zebra and wildebeeste grazing together, interspersed with Thomson's or Grant's gazelle, not far from a herd of impala or a group of kongoni. But we rarely see the 'big cats' together; the tension at the top of the pyramid is much too great for inter-specific sociability.

In the main, the carnivores avoid competition by following different life styles – by being large or small, nocturnal or diurnal, solitary or social. But since several predators are dependent upon prey species common to all, clashes of interest are inevitable.

To us, antelopes and gazelles may seem infinitely abundant in the African grassland ecosystems, but to the large carnivore there are only as many as it can catch; so, in effect, prey animals are rarely over-plentiful and frequently very scarce – especially when migratory herds move away from a territorial carnivore's range. Furthermore, any herbivore is a chore to catch. It is not surprising then, that even the most 'noble' of predators will steal a meal if it has the chance.

The perfunctory displacement of a pair of young **cheetahs** from their wildebeest kill by a maned **lion** (sequence right), is a rare event to witness, but probably happens frequently. Cheetahs' meals are hasty affairs. Other large predators invariably eat the soft belly of their prey first and then consume the rest at leisure. The cheetah, however, begins with the protein-rich muscles of the hind legs, as if expecting to be interrupted at any moment, as indeed were these two within five minutes of making their kill. There was no argument, much less open conflict; the cheetahs abandoned their meal while the lion was bearing down on them, still over 100 yards away. The carcass was almost intact, lacking only the five to ten pounds of rump steak the cheetahs had hurriedly gulped down.

# Competition in the Family

The lion is almost an ecological luxury, and, judging from its arrogance and indolence, we might think it knows this but does not particularly care. The size of this 400 pound cat means that it may select its prey items from the larger species of herbivore such as buffaloes and giraffes. Its size also means that it can steal food from the other predators with virtually no resistance. In Ngorongoro Crater some 80 per cent of its meals are poached from hyaenas, a fact which makes the lion almost redundant in that ecosystem.

Four out of five lions in the Serengeti live in territorial prides. The territories may need to cover 100 square miles in order to contain enough meat to support a score or so of lions. But even so, lions are only sociable because they have to be. Their most successful hunting technique is the stalk or the ambush in which two or more animals co-operate. But once the prey is dead co-operation ceases. The strongest of the pride eats first, and if the prey is a small Thomson's gazelle, as in this picture, there will be nothing left for the others. If prey is scarce then the cubs will die and the young adults will probably leave the pride to seek their fortune elsewhere.

There is an enormous potential for increase among lions. Being the largest and the most powerful of the flesh eaters, they have no natural enemies, and one reason that adult lions are not covered in spots is that they have no-one to fear and hide from. Females are in season every few weeks throughout the year, and produce up to six cubs in each litter. But lion populations never get out of hand, simply because they are so strictly controlled by the numbers of herbivores available. In turn, lions have little effect on the size of the herbivore populations, except in very local situations where the number of herbivores has been critically lowered by some other factor – drought or competition.

The top of the food chain is a lonely, but comfortable niche. The large herbivore diet means the lion lives from one filling, high-protein meal to the next. Often several days are spent over a large carcass, eating and keeping other animals away. One lioness in Tsavo National Park, Kenya, exhibited remarkable persistence over an elephant drought victim. She spent nearly a week lying down repeatedly licking the same spot on the elephant's hide until her cat-tongue finally rasped a biteable hole in the three-quarter inch skin.

Carnivores: Pursuit of food

# A Hierarchy of Strength

At a kill, the cat which was greeting and rubbing heads with her pride a short while before becomes an intolerant individual. Chewing is continually interrupted with low growls, bared-teeth snarls and skirmishes over a choice piece. The more numerous females usually do the killing; the males find it easier to steal the caught food from the females.

In this sequence a **lioness** left a kill to chase off another female, one who had three cubs to feed. The cubs, being the lowest in the hierarchy, are lucky if they get anything. In the Serengeti each year three out of four cubs may die during the dry season, because all of the migratory herbivores have moved a hundred miles North into territories held by other lion prides, defended by males who will fight trespassers to the death.

We are fascinated by these cats because of the functional beauty of their size and strength. We also fear them. One of the most terrifying experiences imaginable is to meet one face-to-face on foot in the African bush – Man, the paragon of animals, instantly reverts two millions years to a defenceless prey. But we can, perhaps, salve our wounded pride with the thought that in the great mill of the grassland ecosystem, the lion accounts for only a minute fraction of the turnover of materials.

Carnivores: Pursuit of food

# The Hunting Clan

It is only recently that the secrets of the **spotted hyaena's** life style have become known. Prior to research done in northern Tanzania, the only picture we had of the hyaena was that of a skulking, scavanging coward – a natural but erroneous conclusion drawn by hunters and tourists who are diurnal and usually looking for more spectacular game. A dedicated research worker, however, is willing to sit through the night until dawn if necessary to see what happens in the few hours when the hyaena is not lying down.

It was always known, for example, that the hyaena is a beast of dubious ancestry, coming somewhere between the mongooses and the cats. Since the time of Aristotle, the hyaena has been considered to be a hermaphrodite. This legend comes from the anatomical oddity of the female's private parts. She sports an outsized clitoris as large as the male's penis, and even has a convincing but empty scrotum. These sham male parts possibly serve some function in the mutual genital-sniffing greeting ceremony, but what exactly the function is remains known only to the hyaenas. As well as their genital ostentation, the females are physically larger than the males and clearly dominant in social encounters.

The main revelation of hyaena research was that, far from being an outsized decomposer, the hyaena is a highly efficient social hunter.

The daytime observation of a crowd of hyaenas waiting for a lion to finish its meal is the last tableau in the common story of a night-time hyaena kill having been stolen by the King of Beasts.

A solitary hyaena is no match for a mother wildebeest protecting her calf or for a stallion zebra protecting his family. But acting in concert the hyaenas almost always succeed. Thus the key to the hyaena's success as a hunter is his sociability. Spotted hyaenas live in large groups dubbed 'clans'. The clans partition out the ecosystem and are fiercely territorial about their propery rights. If one clan chases a wildebeest into the neighbouring clan's territory, the hunt is likely to change into a border skirmish between the two groups. In the centre of the clan territory there is invariably a den dug in the ground where the animals lie up during most of the day, where the young are born and where they shelter whilst the adults are out hunting. The function of their territoriality is to guard a food supply. The function of their sociability is to catch it.

Yet the beauty of the hyaena system is its flexibility. A lone hyaena does not starve, and if the herds of large herbivores have moved away from the clan area, each hyaena can make a living as an opportunist, or indeed, even as a scavenger.

# Competition in the Clan

The **hyaena** can, and eventually does, eat just about anything – it is almost a carnivore version of a goat. But clearly, freshly killed meat is preferred, if available. The massive jaws and musculature can crunch up and eat an entire wildebeest, leaving only the horns. The digestive system processes nearly all forms of protein, so that the droppings are white – almost pure calcium mixed with hair. The hair, incidentally, makes it possible to identify what kinds of prey the animal has been eating.

Although the clan hunts together with beautiful timing and co-operation, at the kill it is every beast for itself. The prey is literally torn to pieces, and we see hyaenas scattering in all directions – one with a hind leg, another with the head, a third dragging off the entrails. Less than fifteen minutes after the kill, all there is left of the wildebeest is a pile of its stomach contents where it died. Out in the darkness we hear the crunching of bones as each hyaena finishes his portion in solitude.

Like many opportunistic predators hyaenas occasionally cache surplus meat. Under water is a common place, since other predators cannot discover the store by smell. Hyaenas have a great affection for water, and certainly do not mind getting wet. The hyaena who hides the meat can find it later by wading around in the general vicinity until he stumbles on it. A hyaena being chased for his prize by other hyaenas may take to the water as a natural refuge. It looks a bit foolish as a hiding technique, since the other hyaenas can clearly see where he is. But it serves to slow down his pursuers if the chase continues out the other side.

Carnivores: Pursuit of food

# The Co-operative Imperative

It would be impossible for a lone jackal to kill an animal larger than itself. But a pack of co-operating **wild dogs,** each not much larger than a jackal, can bring down a wildebeest with remarkable dispatch. In contrast to the hit-or-miss and frequently unsuccessful tactics of lions and hyaenas, wild dogs work as a well-organised team – and rarely fail. They confuse, distract and harass their prey; animals are cut off from the security of their herds and run down in a series of flanking movements which enables fresh dogs from the rear to take over the chase as the lead dogs tire. The prey gets no respite and is quickly exhausted. Thus we see a large wildebeest calf held at bay by two small wild dogs; the rest of the pack will soon join them.

By co-operating like this, wild dogs can kill anything from a hare to a zebra, and hence the absolute amount of food available to their populations is greatly increased. This is the essence of their social hunting and it is crucial to the wild dogs' survival.

The larger, stronger hyaena can strike out on his own occasionally and make a living as a solitary rough-neck scavenger. But the frailer wild dog is bound to his comrades, for without them he will starve or be killed. There is a tempting parallel here with the demands that the slight, fangless and clawless early hominids must have faced – the co-operative imperative.

Every hyaena and lion is essentially out for himself, but the African wild dog is a truly social beast. Their packs are permanent social entities. Their constant togetherness seems to reduce the need for displays of reassurance, appeasement and the suppression of aggression, such as demonstrated by an absentee lion or hyaena rejoining its group.

Instant recognition of who is who and what his status is, may be augmented by the wild dogs' patchwork markings: each animal has a different pattern, and if research workers can learn to tell individuals apart, certainly so can the dogs themselves.

The proportion of females in a wild dog population appears to be unusually low, compared with other large mammals. The more usual preponderance of females is often explained by an argument which points to their importance in breeding. We suggest the dogs' reversal is for the same reason, in the light of their social hunting. Supposing fifty females are needed to maintain a certain dog population. It might take something like ten wildebeeste to feed them and their pups at a time. This in turn would take the efforts of about 100 dogs to pull down the prey. Attacking a large animal can be a dangerous occupation for small carnivores, so to minimise the risk to the child-bearing females, these socially organized animals provide the necessary extra hunters in the form of males. In a hunt where there are, say, two males for every female, any *one* female has clearly half the chance of getting kicked or gored to death. It is easy to hypothesize about the function of excess males, but not so easy to discover the mechanism.

Carnivores: Pursuit of food

# A Hierarchy of Respect

A **wild dog** kill is not pleasant to watch. The prey is often partly eaten before it is dead, the wildebeest bellows and struggles to escape even as its entrails are pulled out and hunks of meat are torn from its rump. Such a death may disturb human sensi-

bilities, but that is irrelevant. A wild dog kill is remarkable, not for its inhuman bloodiness, but for the almost human mutual respect we see among the dogs, and their deference to one another. There is no frenzy, no snapping, squabbling or fighting. The dogs eat in their hierarchical order, each calmly waits its turn, presumably in the knowledge that if there is not enough to go round this time, then the pack will kill again – and again if necessary, until each has had its fill.

Carnivores: Pursuit of food

# The Nimble Opportunists

Given the relative paucity of protein available to carnivores, compared for example to the sheer weight of grass on which grazing herbivores live, it is not surprising that some of the most successful carnivores are the opportunists. Animals like the **jackal**, crow, mongoose, and some **vultures** and bustards take protein wherever they can get it. They will kill or scavenge. Their choice of food items includes most animals smaller than themselves, or larger animals if they are dead. They even eat some vegetable matter. The opportunists literally live by their wits with no special adaptations except perhaps a relatively small size and what we would call intelligence. They cannot afford to maintain too large a body since the source of the next meal is never certain. They occur in large groups or singly, depending on the abundance of food in the area. Their apparent intelligence is born of the need to be exploratory and inquisitive, poking into every niche they can, ready to make use of what they find. The flexibility and ability to adapt to new situations is witnessed by the fact that both jackals and crows can make a living in an environment radically changed by man, even to within the boundaries of our cities.

Carnivores: Pursuit of food

# The Omnivorous Opportunist

From the heights of his mental evolution Man sometimes forgets that he is an animal too; a primate, with distant cousins among the lemurs and monkeys, and closer relations among the apes with whom he shares a common ancestry.

As an animal, Man is governed by natural laws just as much as every other living organism; he and all his works are no less part of the Earth's finite collection of materials than a termite and its mound.

While giraffes evolved long necks, lions sharpened their claws and gazelles learned to sprint, men developed their brains. It is a remarkable tool for living, but makes Man unique only in so far as he can imagine the future as well as consider the present and remember the past. Man is able to analyse cause and predict effect, and the success of his species depends largely on this ability. At this stage he is unlikely to evolve other physical adaptations to aid his survival. The cognitive brain is all he has.

Rightly or wrongly, civilised Man regards himself as the most highly evolved of the species, and surrounded by the trappings of his lifestyle, some individuals may regard the biological fundamentals of life as remote – if not unimportant. But of course, every single piece of all the artifacts of civilisation is nothing more than a part of the Earth. Destroy civilisation, and Man would be left with only the Earth and his brain. Could he survive?

Were such a cataclysmic change of lifestyle to happen suddenly it would probably destroy most of *Homo sapiens*, but there would still be some individuals left to perpetuate the species: foremost among them would be the few remaining tribes of 'hunter-gatherers'. The **Kalahari bushmen**, for example, are able to survive on what the Earth supplies in a basic form. If freed from the strictures of civilisation, these people could leave the desperate retreats to which they have been driven, and roam the continents as once they used to.

Nomadic hunter-gatherers existed at the very roots of human society The biblical phrase 'land of milk and honey' harks back to the small, tightly-knit bands of people for whom perfect contentment was wild honey from the trees and milk in the udders of the nursing antelopes they killed. In the Kalahari today, the Bushmen encounter such luxuries only very infrequently. They live by the bow and arrow, the spear and the snare; they eat plants, insects, rodents, reptiles and birds. Survival depends upon total co-operation among band members. The social customs they have evolved obviate the ownership of property, and complicated allocation procedures ensure that everyone gets a share of a kill or foraging expedition. For instance, the old man who made the arrow is entitled to a part of the antelope it killed, and the woman who supplied the bag gets a proportion of the berries it brings back, no matter who filled it.

Each band has mutually recognised rights over areas of land, though intermarriage brings with it shared rights. This is particularly important at the height of the dry season when food is very scarce, and the water available is in moisture-bearing roots that must be dug from underground with sticks and the brain's ally – the hand.

It is an extremely hard life in which there is no choice – only necessity. The Bushmen are an integral part of the ecosystem in which they live: they bring nothing to the desert and they take nothing from it. But, while they do have some minor physical adaptations peculiar to themselves, they live by the basic tool of the species: the brain tells how to kill the antelope and the brain suggests where to look for the succulent roots.

# Seizing an Opportunity

Sources of nourishment are found everywhere in varying degrees of availability. The front end of most animals is usually designed to deal with particular sources of food, and feeding involves the appropriate use of particular tools. Body form and behaviour are always functionally and operationally inseparable: the owl's eye and night hunting; the heron's bill and fish stabbing; the hyaena's jaw muscles and bone-crushing; the lion's claws and prey grabbing; the eagle's beak and flesh tearing – the part and the use evolved together. The propensity to use the right part of the body in the correct way is innate, inborn, but early attempts by young animals are clumsy. Efficient use must be perfected through play and practice and the rewards and punishments meted out by the environment which result in learning.

Ostrich eggs are attractive sources of nourishment but relatively rare, so the **Jackal** (lower right) and the **Tawny Eagle** (upper right) have not evolved the wherewithal nor have they learned to open them. But the **Egyptian Vulture** (left) has: it throws stones at the eggs.

So intimate are body and behaviour that we are not surprised when a tse-tse fly plunges his proboscis into us, or when an eagle tears off pieces of prey. But when an animal uses a part of the inanimate environment to exploit a food source – like the Egyptian vulture throwing a stone at an ostrich egg – we are amazed. There is no reason to explain 'tool using' by invoking a superior intelligence, any more than such an intelligence is necessary to explain the woodpecker's use of its bill to hammer holes in trees. Stones are ubiquitous and frequent parts of the environment – even more so than birds' beaks. The components of the stone-throwing movements – picking up something in the beak, lifting the head up and bringing it forcefully down as in an attack or a strong peck – are not unusual. Given the 'tools' in number, the movements already evolved, and a reward equivalent to two dozen hens' eggs, the stone-throwing trick is not so surprising. We do not see more tool-using because most species have developed built-in equipment to exploit their most frequent food sources. The Egyptian vulture probably has not evolved a large ostrich-egg cracking bill swung by enormous neck muscles, because ostrich eggs are not available all the year round: foods which put a premium on a slender bill are. On the other hand, when ostrich eggs are in season, the amount of protein they provide makes it worthwhile for at least one member of the community to do something about them.

Decomposers: Pursuit of food

# Dependence on Death

At the very top of the pyramid of life perch the different kinds of **vulture**. We might think of them as predators who happen to prey on animals that are dead. The wooded grasslands provide a large quantity of vulture 'prey'. Eventually all species fall into this food class. A live Thomson's gazelle and a live zebra provide a predator with very different prey catching problems. But once two such animals die, they become effectively one 'prey species' for vultures.

# Specialist Scavengers

Although all animals eventually die, carcasses tend to be much more scattered in time and space than herds of live animals. Thus, for a group of **vultures** around a carcass, the essential resource at that particular time and place is limited. As one would expect, this situation produces competition, which in turn provides a selective pressure for the birds to find different ways of attacking the resource: the result is that there are six different species of vulture – the large **lappet-faced** (flying) and white-headed; the medium-sized white-backed and **Ruppell's-griffon** (foreground); and the small hooded and Egyptian.

At first sight vultures on a carcass appear to be involved in a fierce free-for-all struggle for food; but there is much more order than might be supposed. Fights are predominantly between members of the same species. Different species arrive on the kill at different times in a consistent order; they utilize different parts of the carcass and are specially equipped to deal with those parts; they have differences in carcass spotting abilities, feeding behaviour, nesting habits and levels of aggression. Thus, even though they might bicker over space at a kill, each species occupies a distinct niche, which allows them to co-exist.

The classical image of the vulture skulking on a dead branch and casting a baleful eye over the ecosystem is apt enough – they are continually watching and waiting for someone to die. They range over thousands of square miles. Meals, as we have said, are relatively scarce, so the searching vulture conserves fuel by soaring. It can travel many miles without a wing stroke, by cannily choosing a route through thermal updraughts and air rising on the windward side of hill slopes. So efficient is this low-budget travelling, that they can afford to maintain the large, strong bodies necessary for competition at a carcass. And so efficient is their searching, that if you lie down to have a nap in the grass, you may be shortly awakened by a vulture landing next to you.

No matter how spectacular the herbivore, how ferocious the carnivore, all eventually yield their material to the decomposers.

Any relationship between a vulture and a **dung beetle** may seem decidedly tenuous, but they both – along with a variety of species of flies, crickets, moths and bacteria – help in the disassembly of organic matter and so speed up the return to the soil of essential chemical elements.

To serve this end and, of course to maintain themselves, most of the decomposers can fly. It is a matter of minutes before the first decomposer finds a fresh carcass or dung pile. The vultures will be there before rigor mortis sets in, the flies within seconds of the animal excreting. They all work with incredible speed. In the wet season, when dung beetles are most numerous, a pile of elephant dung, all twenty pounds of it, is worked into balls and rolled away, almost before it cools. The beetle (Coprinae) deposits eggs in the ball, which it then buries, three feet underground. Thus manure is injected directly into the earth. When the eggs hatch the developing larvae feed on the dung, taking the process of decomposition a stage further.

# II

# LAKES & RIVERS

# The Accumulation of Water

Twenty million years ago in Africa the earth gave a lurch, volcanoes erupted and two of the dozen or so great plates which make up the surface began to drift apart. The result was **the Great Rift Valley,** 4,500 miles long from the Dead Sea to the Zambesi Valley and 50 miles wide just west of Nairobi. The faults and terraces of the rift are still as clear as when they were formed. In the bottom of the several arms of this valley lie the most important East African lakes, the surfaces of which still occasionally ripple with volcanic tremors.

The ecological character of the lakes and rivers is determined by mountains. The water that falls or condenses on them is pure, distilled from seas and lakes. The greater part of it, moving down a well-covered slope, actually flows below the ground until it reaches the junction of two slopes, where it surfaces as a spring – the beginning of a river. Whilst percolating through the soil and over rocks, the water takes on the chemical character of the mountain itself.

Water which flows from heaved-up granitic rocks like the Ruwenzori in Uganda, the Usambaras in Tanzania or the Matthews Range in Kenya, for example, tends to be acidic. Moreover, the granite and sandstone particles filter the water to a state of crystal clearness, almost too pure for the tastes of life.

On the other hand, many of East Africa's mountains are not-so-ancient volcanic ranges, composed of alkaline lavas. Consequently, the water which runs from them carries the basic salts of the rocks into the lakes where they accumulate, for many of the Rift Valley lakes are closed basins with no outlets, like Lake Magadi, seen here.

The transition from the land to the water is as abrupt as the edge of a lake. Living in water poses problems unknown to organisms on land. For example, getting oxygen, orientating in three dimensions without a horizon, feeding on primary production which floats unattached, all require special adaptations. We say this, of course, from our terrestrial perspective. Evolution worked the other way round, from the water to the land, and most of the life we encounter in the grasslands or the forests is supported by special adaptations to escape the medium in which life began. Despite the obvious difference between aquatic and terrestrial ecosystems, we shall see that the same basic ecological rules apply to the entire spectrum of life in the water, which ranges from microscopic plants to herbivorous hippopotami and carnivorous crocodiles.

**Primary production: Controlling factors**

# The Movement of Water

The effect of mountain slopes on lake and river water composition is not only chemical. There are physical effects as well. Water which rushes down mountain streams or thunders over cataracts like **Kabalega Falls**, churns and froths and takes oxygen directly from the air. This process may be a matter of life or death for aquatic animals. Still lake water receives its oxygen from the quieter process of plant transpiration.

Erosion, or more euphemistically 'redistribution of soil', is another physical process. It occurs everywhere to a greater or lesser degree, and even on a non-abused, well-covered slope some soil gets transported into the drainage lines. Excess water first trickles down the hill, then as it joins with other rivulets it gathers power, cuts into the sides of the stream bed and carries away soil. It rolls out stones and boulders chipping off the sharp edges and making round pebbles of them. Streams run into rivers, rivers to the sea, or into swamps, or into lakes.

There is always some movement of this sort going on in the mountain rivers, which swell enormously during the floods that follow heavy rain. In fact, very few East African rivers are perennial, like the Nile, the Tana or the Great Ruaha, which flow all the year round. Most are seasonal, flowing briefly during the rains and dry the rest of the year. Water is present, though, below the dry beds, but it takes the ingenuity of man or elephants to dig wells to get at it. The Kerio in Northern Kenya is an extremely erratic river, which only enjoys enough rain actually to flow into Lake Turkana once every few years. Most of the time it is a 'sand river' more useful as a track than a watering place.

Where the rivers run into lakes the soil and débris they carry is deposited on the bottom. As the velocity of the water diminishes, first the heavier stones and gravels are deposited, and then the smaller particles of soil. Finally the fine silts fall slowly to the bottom, or may even remain in a colloidal suspension, giving a particular colour to the lake. Lake Michaelson is still green from the fine silt particles carried down by the glaciers of Mount Kenya many millions of years ago.

# Flood and Drought

The processes of solution, erosion and deposition are continuous and relatively slow, except in cases of deforestation. They vary in effect from day to day, from year to year and from age to age. The onset of the rainy season can be sudden and result in quick floods carrying silt from parched and eroded land. These floods rush into the lakes and may, perhaps, drown or wash away the nests of birds, initiate the migratory movements of spawning fish or, over a longer time, bring about a drastic change in the lakeshore habitat.

**Lake Nakuru** rose dramatically following the 1961 floods, and has dried out four times since 1951. From their girth we estimate that these stumps of yellow barked acacias were mature trees forty to fifty years old when the rising lake waters drowned them. Since we know from research in Amboseli National Park that such trees cannot tolerate highly alkaline ground water, it follows that the lake's water table must have been considerably lower than it is now for the entire fifty years the trees were growing. Further testimony to the lake's inconstancy.

Ironically, the trees died not so much from an excess of water as from a shortage of it. This is because in the process of osmosis water travels across organic membranes (such as the cell walls of root hairs) towards the side of highest salt concentration, so the trees actually lost more water to the alkaline salt solution of the lake than they could extract from it.

Clearly, such fluctuations are related to climatic trends. Even the investigator who discovered that the level of Lake Victoria changes in an eleven-year cycle directly parallel to sun-spot activity would have to admit that the effect must somehow be one of sun-spots on rainfall.

Through the much longer timespan of thousands and millions of years, life in the lakes has changed profoundly, though gradually. Lake Naivasha, for instance, was once joined through rivers and swamps to another large lake at Olorgesaile. Around these waters our earliest ancestors lived. That lake is now gone, and, though it may seem difficult to believe when you see its deep clear waters, Lake Naivasha will eventually disappear too, filled with silts, rocks and soils of the higher ground and of the mountains.

So the character of a particular African lake may be quite different from its nearest neighbouring lake, depending on the character of the surrounding land. The perversity of rainfall in semi-arid regions also means that a particular lake ecosystem may not be stable for long. In ten or twenty years the lakes we observe today could be quite different, in size, in numbers and types of species they support – some may even be gone completely. Nevertheless, despite their ephemeral nature, the laws governing their functions are the same as those that pertain to the grasslands and the forests. For the purpose of demonstrating them at work in the water we will concentrate on the ecosystem of Lake Nakuru in Kenya, where life itself is prolific, but the actual number of life forms amazingly small.

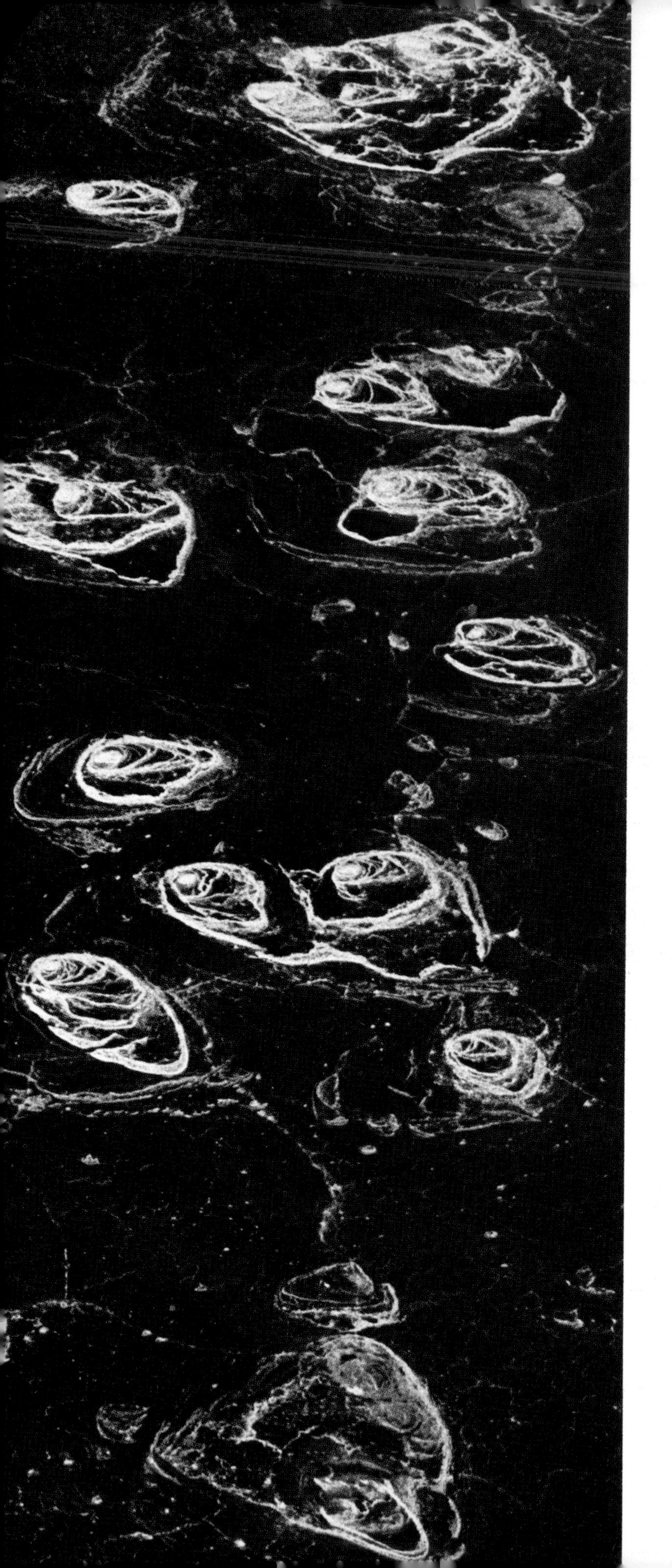



# A Special Solution

Several of the Rift Valley lakes are called by the Maasai 'magadi', which simply means **soda**. In these closed lake basins, dissolved materials, mainly sodium carbonate and bicarbonate, build up in concentration and produce a fertile, if alkaline, environment. When the lake level falls – because of a prolonged rainless period or simply from the evaporation of the sun which volatizes thousands of gallons a day – large deposits of salt crystals precipitate on the shore in sheets of white, brown and red. The crystalline patterns of the precipitated salts are as striking from the air (left); as they are under the microscope (right). In solution they create an unusual chemical environment, one which could kill a thirsty wildebeest but which supports a remarkable profusion of life.

The alkaline lakes are extremely rich in primary production, as we shall see, but they provide a physiological environment so alien to most life forms that only a few have evolved to cope. Leave your hand in Lake Magadi for an hour and, apart from getting soda burns, you will lose water through the skin in a vain osmotic attempt to dilute the lake to a reasonable ionic concentration. The 5 inch cichlid fish, *Tilapia grahami*, does not shrivel up and die in Magadi because it can tolerate chemical concentrations within its body almost equal to those of the water. It survives in a hostile environment by becoming part of it. It can also tolerate water temperatures up to 106° F., such as occur in the vicinity of the hot volcanic springs which bubble out of the rift valley floor and feed Lake Magadi.

Once an organism has adapted to the extremes of an alkaline lake it is likely to do extremely well. It has been estimated, for instance, that the tilapia population of Lake Nakuru weighs 880 tons, the same as 300 good-sized elephants. The fish take about 70 tons of algae a day from their 15 square mile aquatic pasture; a weight of greenstuff that would actually feed 450 elephants, though they could only find that much on a daily basis by ranging across far more than 15 square miles of wooded grassland. These facts illustrate not only the potential richness of a tropical lake and its concentration of nutrients compared with other systems, but also the different nutritional requirements of different sized animals.

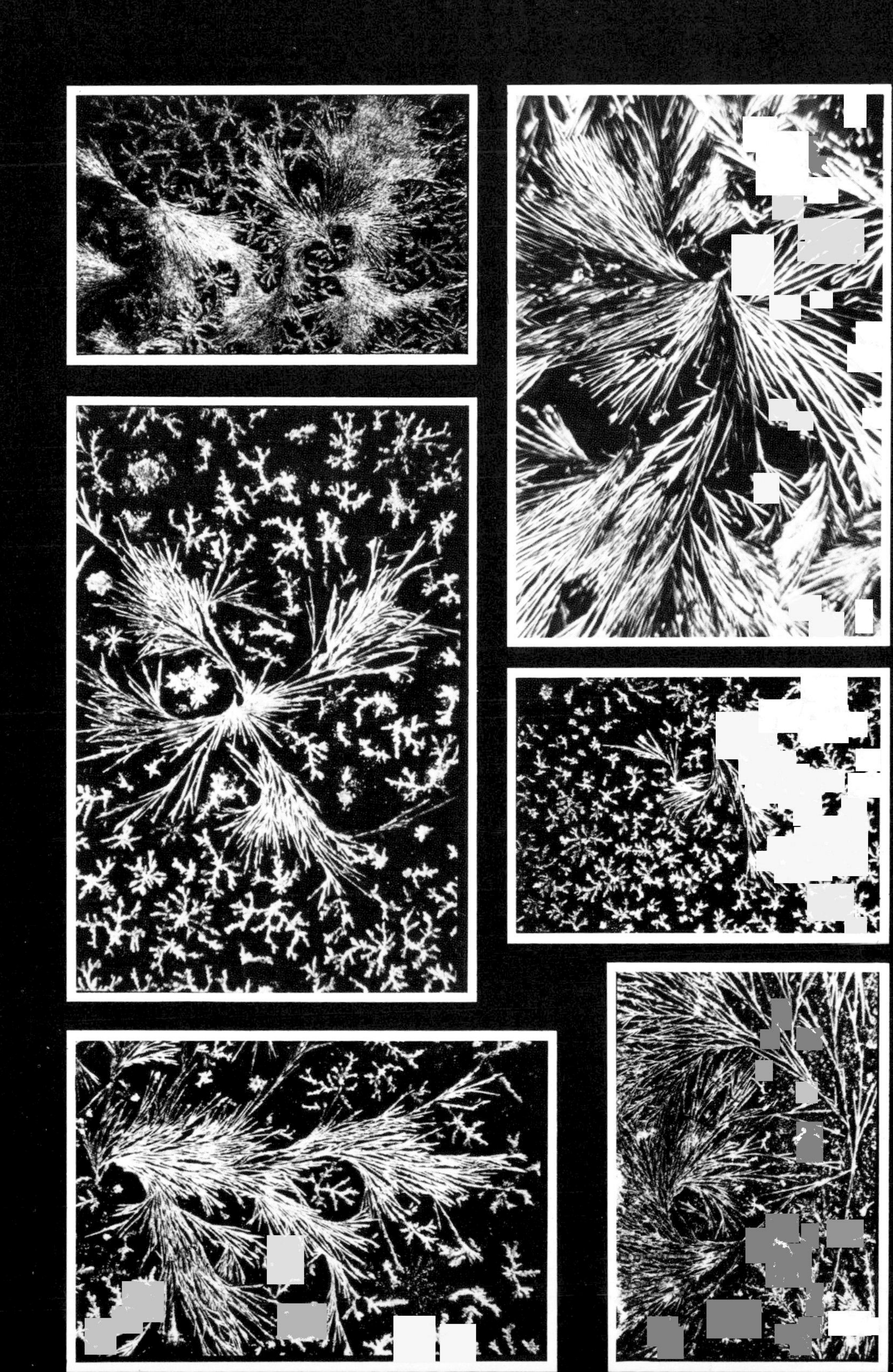

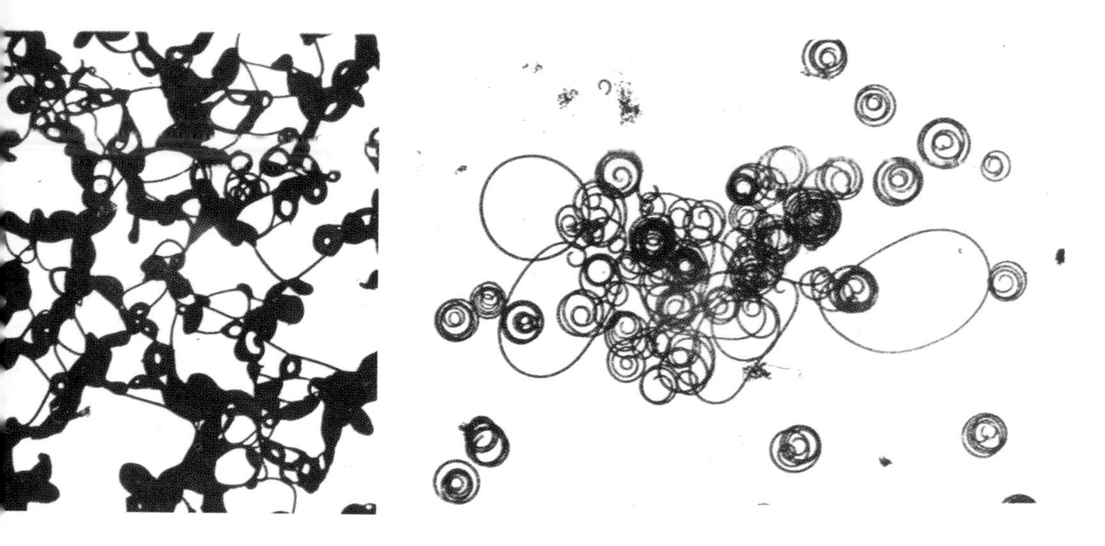

Primary production: Basic food sources

# The Microscopic Algae

Alga is the most important primary production of the lake ecosystem. Ubiquitous and successful, the numerous species of single-celled organisms should not be thought of simply as some form of 'dissolved grass'. Many exist as discrete cells; some clump together in colonies which have the form of proper plants; all have some sort of food chemistry catalyst, usually a form of chlorophyll; others are simple consumers in the sense that they gobble up large inorganic molecules. They reproduce sexually or asexually; they have minute internal organs, organelles, which carry out such functions as excretion and digestion; some can see light, and most can move through their watery medium. They are animate plants, or proto-animals, depending on how you look at them. In terms of evolution, they echo life's first form on earth.

The water of Lake Nakuru is thick, slimy and green. Your hand is invisible six inches below the surface, obscured by the mass of algae suspended in the water. Ninety-five per cent of the cells are in chains belonging to the **blue-green alga, spirulina**, and are so small that thousands of them laid end to end would be about as long as their proper scientific name: *Oscillatoria* (*Spirulina*, *Arthrospira*) *platensis* var. minor Rich. The whole lake is a veritable soup of nourishing vegetation, an expanse of extremely productive pasture. In an algal 'bloom', the weight of *Spirulina* in the fifteen square mile lake may be of the order of 200,000 tons. That is equivalent to the productivity of over one hundred square miles of good grassland.

The algae occurs in such enormous quantities because of the combination of strong sunlight, high temperature and high carbonate and phosphate concentration in the shallow soda lakes. The sofa-spring shaped *Spirolina* is a green plant, and needs sunlight. But so opaque are the waters of Nakuru, that only the top ten inches receive enough light for photosynthesis. Yet the lake is virtually filled with living algae. The reason for this is that every day the shallow waters of Nakuru are thoroughly stirred by wind and wave action, which allows all of the *Spirulina* access to some light.

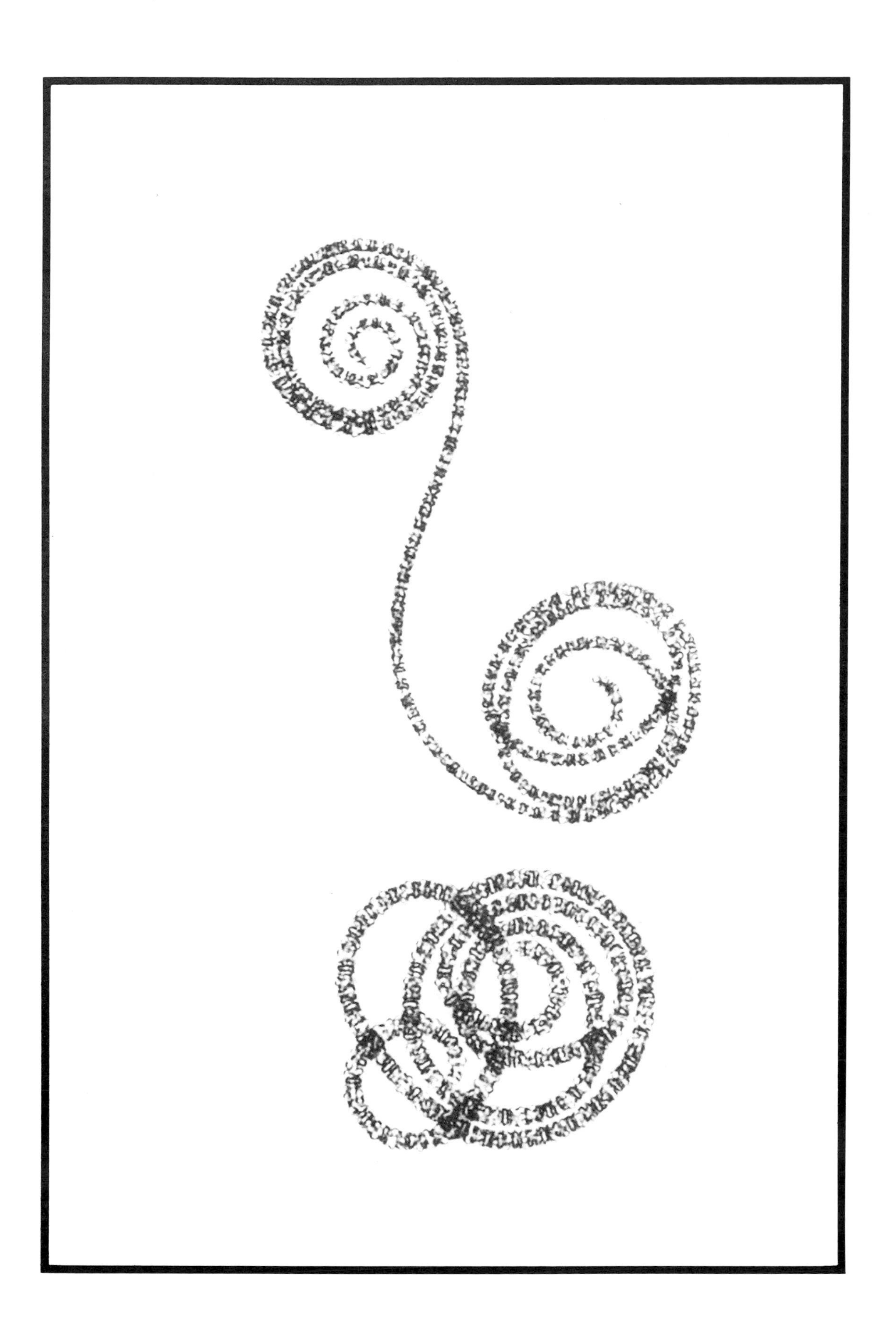

# Enriching the Water

The **hippopotamus** is an invader of the water ecosystem. But since it derives no nourishment from the aquatic food chains, and simply uses the water as a day-time refuge from the sun and from predators of its young, it is a benign guest. In fact its presence is decidedly beneficial. At night the hippos leave the lakes and rivers and wander along well defined avenues up to seven miles inland to graze. The materials they ingest are imported into the lakes when the beasts lumber back in the morning, have a drink and excrete. Thus they speed up replenishment of the aquatic nutrient pool which, without the hippos, would be solely dependent upon the painfully slow process of the weathering of the rocks.

An adult hippopotamus may weigh two tons and therefore consumes a prodigious amount of grass around the lake periphery. Its contribution of materials to lake ecosystems may in some cases provide the nutritional basis for an aquatic pyramid of life.

Although the hippo brings a considerable amount of nutrients to the lake, the chance that those materials will subsequently find themselves moved back across the shoreline during their progress through the food chain is fairly low. The chance is greater that they will travel from trophic level to level within the lake. Perhaps in the long run the hippos' rent is offset by those birds who feed from the lake but then fly off to excrete or die elsewhere.

The point is, the materials move round within the lake at a higher rate than they move in and out. This defines the ecosystem. Similarly, the odd wildebeest may wander out of Ngorongoro crater and get eaten by a Serengeti lion, but most see out their lives in the crater. Or, a forest-dwelling bull elephant may come down from Kilimanjaro and succumb in the swamps of the Amboseli basin. However, this sort of event is relatively infrequent compared to birth, feeding and death within the entity of the ecosystem. The hippopotamus is one of the few mammals that makes a habit of regularly transgressing ecosystem boundaries.

**Herbivores: Pursuit of food**

# The Micro-Grazers

A microscopic flora encourages a tiny fauna. Under a microscope in a drop of lake water we observe minute **protozoans**, single-celled herbivores (left) filled with recently ingested *Spirulina* coils.

They are much like the algae they eat except, lacking chlorophyll, they cannot produce their own food. The protozoan herbivores absorb the tiny plants through their cell walls. The algae are broken down into their chemical parts which nourish the protozoa. Waste products from this process accumulate in a bubble which eventually pops through the cell wall back into the lake.

**Rotifers** (right) are not much bigger, but are slightly more sophisticated than the protozoa. They are curious beasts, rather like primitive mechanical worms. They have a row of constantly moving hairs at their front ends which give the impression of two little wheels spinning eddies of water into the mouth cavity. Particles of food, occasionally bits of *Spirulina*, will also be swept in. Eggs are produced mainly by parthenogenesis, that is, without recourse to sex. In fact,

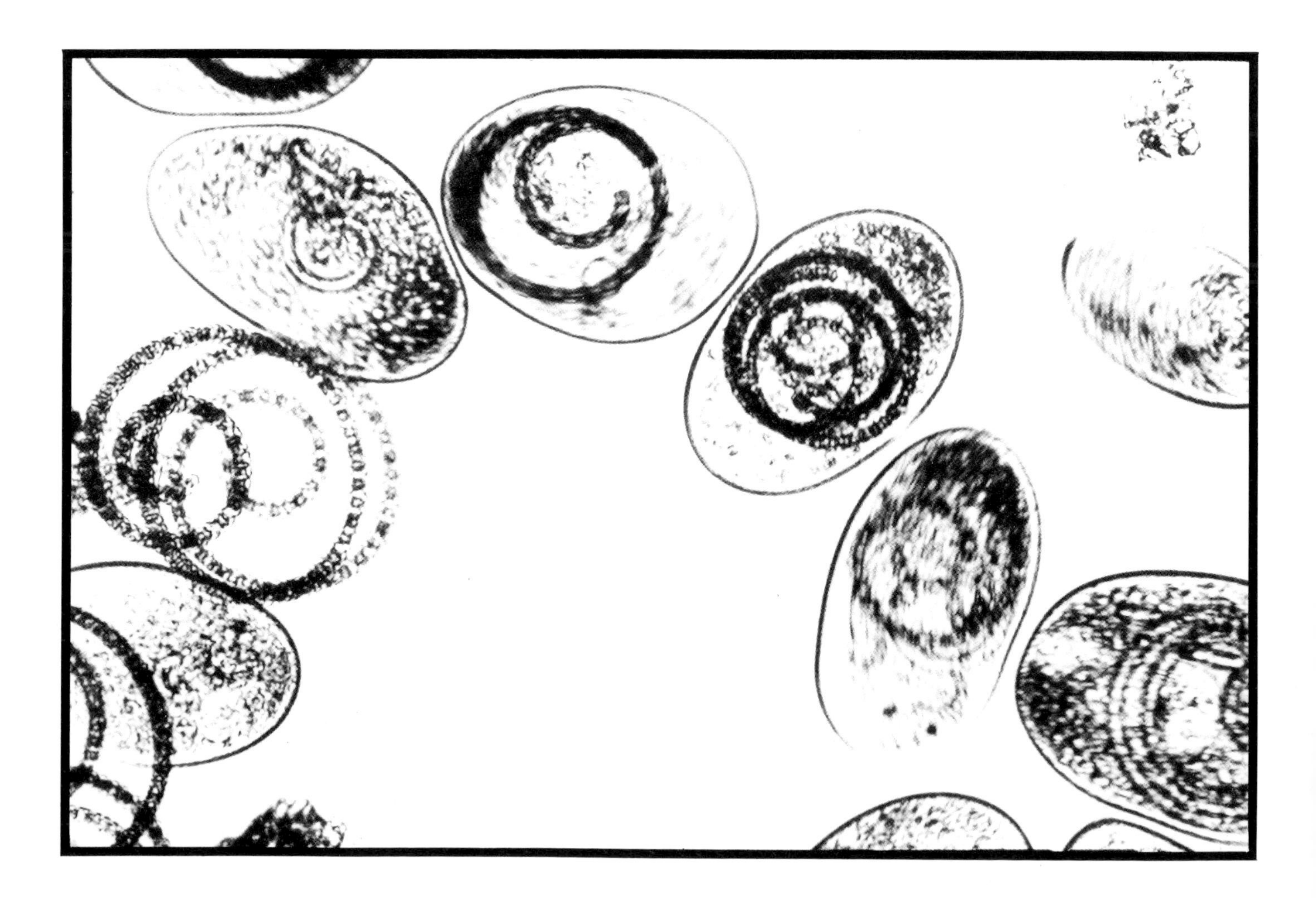

males seem to be completely absent in some species. One type of egg is thin-walled and develops into a miniature adult soon after it pops through the mother's body wall. The other type stays attached to the adult without hatching and seems to be reserved for hard times, when the water is drying up. These eggs have a thicker wall and are remarkably tough. They may be transported about on the legs of wading birds or even blown away by the wind in the dust that was the lake bottom. Some have even been collected floating like tiny time capsules in the débris of the upper atmosphere. This has led to the somewhat fanciful suggestion that rotifers are earth's first space visitors.

The little crustacean *Lovenula* is also a primitive lake herbivore. Not much is known about them except they live only twenty two days, feed mainly on *Spirulina*, and, along with the protozoans and rotifers, make up an impressive microherbivore biomass of well over 2,000 tons in Lake Nakuru. This is roughly equivalent to the weight of the 13,000 wildebeeste on the Athi-Kapiti plains near Nairobi.

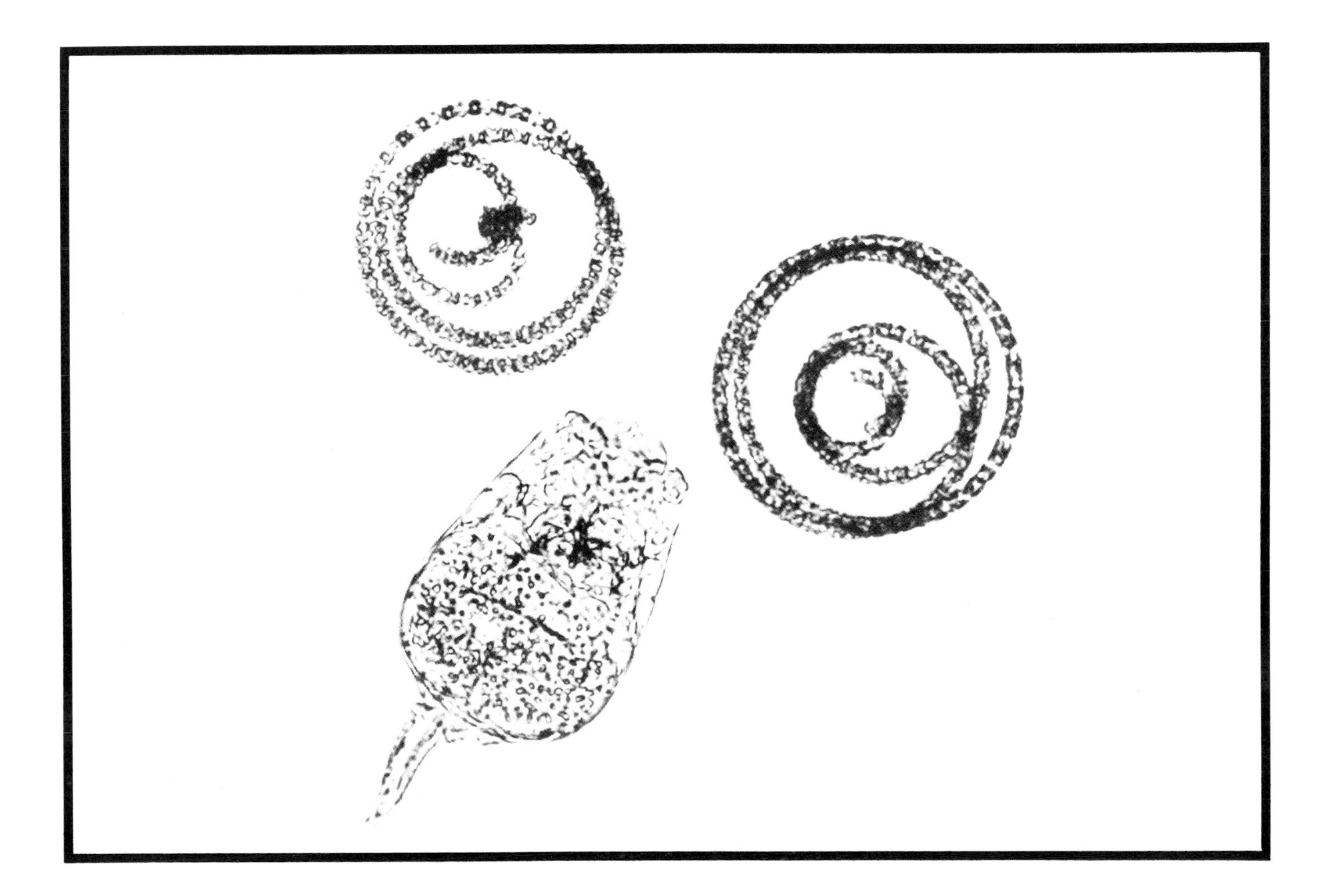

# The Specialist

Specialization usually produces a marvel of design and architecture. The **lesser flamingo** specializes on the blue-green algae like *Spirulina*. Its long legs allow it to wade and feed in water up to three feet deep. Frequently algal-rich waters are shallower than this, so a long neck is essential for reaching the surface. A duck, swimming on the surface, can feed on algae and other small floating plants with its head in an upright position. If the flamingo were to keep its head upright, it would need an extra kink in the neck. It solves the problem by feeding with its head upside down.

Flamingoes harvest algae by 'grazing' on the top inch or so of the water. Inside their bills they have a remarkable filtering mechanism. The upper mandible (detail) is triangular and fits tightly into the lower bill when it is closed. The inner surfaces of both are covered with fine hairy processes, arranged in rows, about 100 to the inch. The thick and fleshy tongue is confined to a tubular groove in the lower mandible where it moves back and forth like a piston in a pump. As the piston retracts water is sucked in over the filter hairs; as the water is forced out again, the hairs erect under the pressure and catch the algae which collects in masses on the downy inside of the mandibles. These masses are rolled onto the tongue and down the gullet in a continuous stream propelled by the pumping motion of the tongue.

When feeding, a flamingo submerges its bill in the water – upside down – and a rapid pulse of water can be seen booming out of the beak, about 17 strokes a second. The lower mandible, which is uppermost in the feeding position, is cellular and filled with air spaces. It is so light, that it will float like a cork if detached from the bird's head. This buoyancy allows flamingoes to feed in choppy water without having continually to use energy in adjusting the height of their heads to compensate for waves.

Thus flamingoes are an extreme example of specialization in cropping primary production. Since they utilise only the top two inches of algae and do not feed over the entire lake, it is clear that they take only a minute fraction of the algae available. Yet even so, it has been calculated that an average population of 300,000 removes about 180 tons of algae daily from lake Nakuru.

# Dangers of Specialisation

Specialists like the **lesser flamingo** are evolutionary gamblers who have put most of their survival bets on the abundance of one type of food. When that type is in plenty, the specialists thrive and do far better than other consumers who are not specifically designed for the job. Thus flamingoes on Lake Nakuru have on occasion reached a staggering 1.5 million birds. All is well as long as the food source is present. In February 1974 however, the culmination of a chain of external events drastically affected the *Spirulina* population. A long dry period reduced the lake level and consequently increased the alkaline concentration in the water. It was too high for the *Spirulina* and two-thirds of them died.

The flamingoes had two choices – to move and look for blue-green algae elsewhere, or to die. They made attempts at falling back on an alternate food source, such as the green algae *Clamydomonas* which took the place of *Spirulina*. But specialization left the birds at a disadvantage. The green algae are smaller than the blue-green, and the flamingoes' precision-built filters let too many escape. Thus the flamingoes did not get enough to eat even when the waters were as turbid with green algae as they had been with blue-greens a few weeks before. They were forced to change their feeding habits altogether and spend long hours sieving the lake-shore mud for single-celled plants called diatoms. These are a limited food source compared to the profusion of an algal bloom, and can only support a fraction of the algae-fed flamingoes. In a few weeks the flamingoes of Nakuru all but disappeared – some died of starvation, most migrated.

There are three interesting questions connected with the flamingoes' dilemma: how did they know when to go, how did they know where to go, and where did they in fact go? Doubtless they decided when to go by reacting to their stomachs. There came a point in the decline of the algae when each flamingo was not getting enough. Awareness of this, which we would call hunger, resulted in a restlessness which caused the birds to fly away from the lake. Any body of water would prove attractive to the migrating flamingoes. They would land and test the algal pastures or the diatom content of the mud. If sufficient, they would stay. If not, the same forces would compel them to move on. They clearly dispersed to lakes up and down the rift valley – north as far as Lake Turkana on the Ethiopian-Kenyan border, and south, some authorities hold, as far as Botswana, some 2,000 miles away.

But it is not all that easy to explain their movements, for within a few weeks of a recurrent algal bloom in Nakuru many of the birds were back. So the specialist either needs the ability to reproduce very quickly in order to take advantage of gluts of his special fare, or the ability to move and seek out new food sources with an uncanny timing and awareness.

Herbivores: Reproduction

# A Place to Mate

In the warm alkaline waters and abundant food supply of the lakes, **tilapia,** *Tilapia grahami*, breed all year round. They are fiercely territorial – but only in the mornings. At sun-up the males square off over the centres of their 18 inch diameter territories. Other males who wander into the territories receive open-mouthed threats which feature the conspicuous white lips. Adjacent territorial males gape across the boundaries and probably bombard each other with invisible water pressure blows by a sub-marine equivalent of puffing. Transient males bounce through the area from one territorial encounter to the next, whereas schools of females and young sweep through the breeding grounds leaving a wake of posturing territory holders, who suddenly switch to behaviour designed to attract gravid females rather than repulse invading males.

At eleven o'clock, they all knock off for lunch, leave the breeding grounds more or less in company and seek out algal grazing areas. It sounds curiously sporting – a jolly good morning of territorial aggression and a spot of mating, followed by a light but nourishing lunch, and an afternoon of social idling, snacking and dodging pelicans or darters. The next morning the same males will be back in their territories threatening their neighbours.

# A Place to Brood

We saw in the wooded grassland that some territorial animals like the impala defend a space in order to protect a food supply. *Tilapia grahami*'s little territory clearly has some other function, since it is too small to provide enough algal food and is usually in a part of the lake where there is little algae growing, such as in clear water near a spring. The territories then, appear to serve mainly as a meeting and mating place.

When the male **tilapia** (upper left) is not repelling intruders he spends his time tidying up his 'breeding pit' in the centre of his territory. These 3 inch depressions in the lake floor are carefully excavated by picking up sand particles in the mouth and spitting them over the rim. We would guess that they are constructed in concentric terraces to be visually conspicuous to passing females.

If a female leaves a school to inspect a pit, the male courts her with an eager side-on display (lower left) which catches the light, showing the female the splendour of his breeding colours, and at the same time points his otherwise threatening mouth in a neutral direction. She is likely to be receptive if she was interested in the pit in the first place, and mating occurs over the breeding pit. Territorial neighbours look on, but do not interfere. They have learned not to trespass to the centre of the territory where the holder's reaction is most violent.

*Tilapia grahami* is one of the rare species of fish which is a mouth brooder. Soon after the eggs are laid, the female takes them up in her mouth. It is not known whether egg and sperm meet in the open water or in the female's mouth. In either case, the brief period between laying and picking up is one of potential risk, since other *Tilapia* are cannibalistic on the eggs. Another major function of the territory, then, is to provide a relatively safe venue for the few moments when the eggs are in open water between one end of the female and the other.

When the eggs hatch, the fry stay in the female's mouth until they are about half an inch long (right). During the two-week period, the female shows a remarkable restraint and sensitivity, for she feeds all the while without swallowing her offspring, at least not too many of them. The ethologist Konrad Lorenz once observed a mouth-brooding mother of another species faced with a particularly tempting but large morsel of food. She paused, spat out her fry, gobbled up the food, and took up the young again in a flash.

# The Changing Amphibian

The frog is an ambivalent creature. Not only does it live a bit like a fish and a bit like a reptile, but it is also both a herbivore and a carnivore, depending on how old it is.

The first stage of its life, after emerging from the gelatinous egg, is the schoolchild's friend, the tadpole. Fully aquatic, with gills and a fish-like shape, the tadpole feeds mainly on vegetation and detritus. Through a gradual metamorphosis, which is nearly as complete but less dramatic than the butterfly's, legs sprout where before there were none, the tail recedes, lungs develop and the tadpole makes more frequent trips to the surface to gulp air. Until one day, it crawls out of the water and into the next higher trophic level, as a carnivorous adult frog.

In the chronicle of animal evolution there are a handful of adaptive events of such importance that they literally changed the face of animal life on earth. Lungs, wings, and the cognitive brain centres of early man were crucial innovations. All these adaptations have the quality of opening the ecological door to completely new niches, for example, to unexploited food sources. The development of lungs allowed Devonian amphibians of 300 million years ago to use oxygen directly from the air. Unlike their fish relatives they could then hunt invertebrate prey species on land. But the amphibians were and still are bound to return to the water to breed, because their spawn must be kept wet to live.

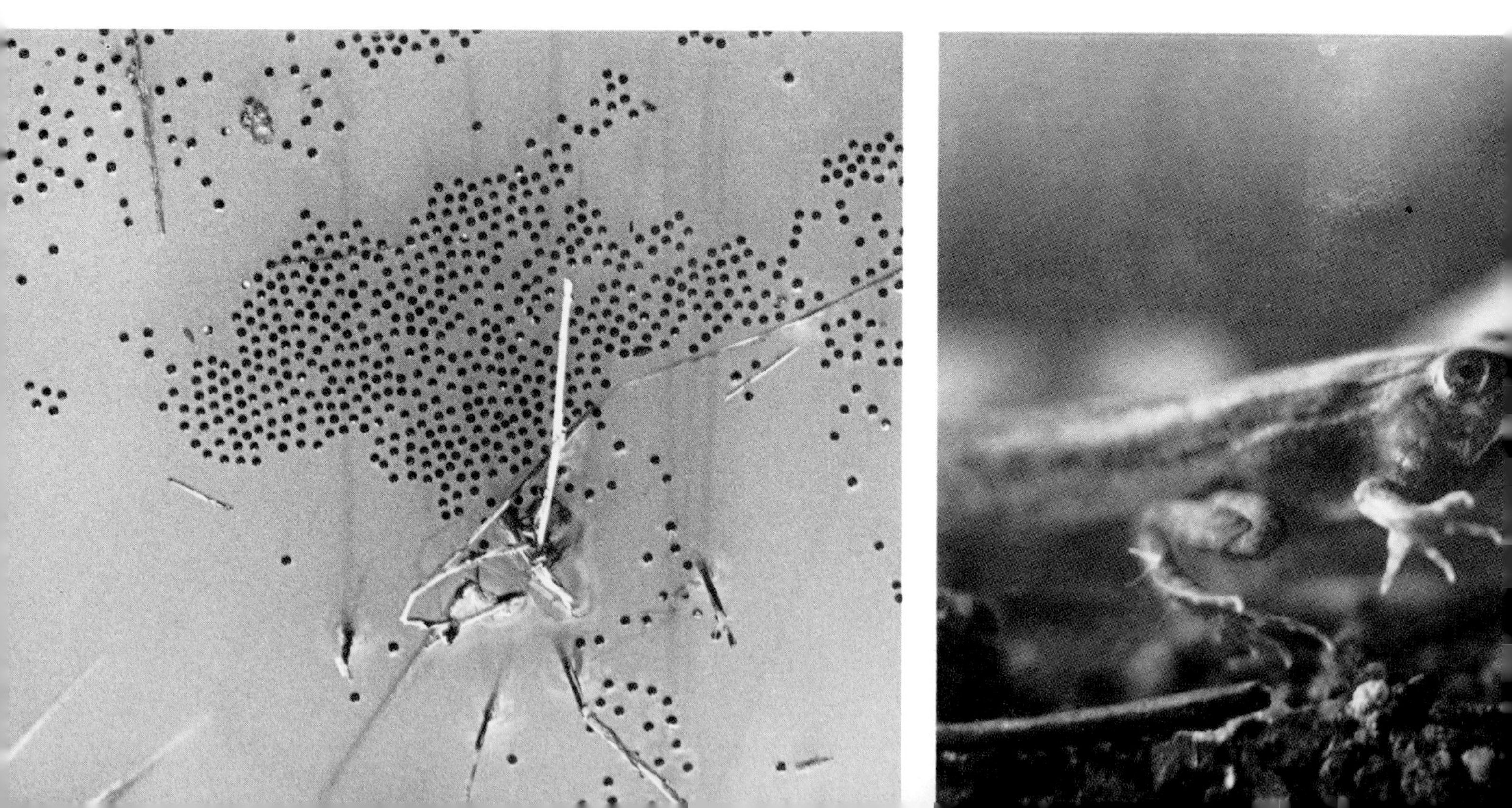

250 million years ago, perhaps during a period of alternate floods and desiccations, it became advantageous for some amphibians to develop another major innovation: an egg which would not dry out even though it suddenly found itself on dry land. A simple but ingenious folding and fusing of an embryonic membrane enclosed the developing embryo in a bag of water – the amniotic sac. Adding layers of calcium on the outside ensured the water would not seep out and evaporate before hatching. In this way the early reptiles shook off the bonds of the aquatic environment by means of an egg which contained its own diminutive frog pond within.

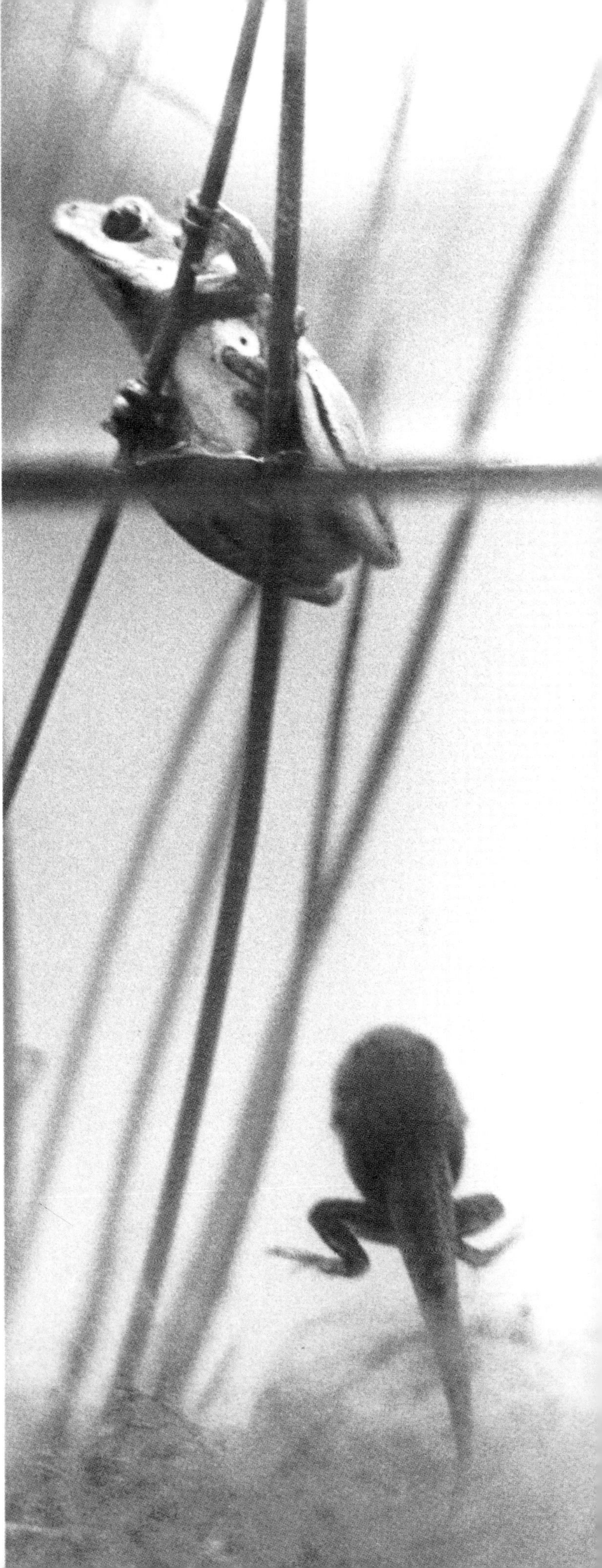

# The Aquatic Dinosaur

Only a handful of reptiles in the world are not carnivores. The rest, like the **crocodile**, need the high protein and energy food which the bodies of primary consumers provide. One hundred and fifty million years ago, during the 'age of reptiles', the earth teemed with several times as many genera as there are today. With a cooling in the climate, the soft swamp vegetation of the Jurassic period diminished and so did the herbivorous reptiles. Although the fall of the reptilian rulers was a complex ecological event, one reason was that their jaws and teeth could not cope with more robust terrestrial vegetation. Curiously, reptile jaw structure and musculature have never progressed very far past the snapping stage, much less to grinding. Even today, the crocodile's trap-like jaws are hinged only to open and shut, the teeth designed to grip rather than do any intricate dissection. The toothed trap snaps over the whole fish, or over an exposed part of a large prey item such as a misplaced terrestrial leg. It is held fast, and the body of the crocodile spun around by a corkscrew snap of the massive tail. Chunks of meat, bone and sinew are thus twisted off like pieces of toffee and swallowed whole.

Despite the grisly way of dealing with prey, crocodile mothers are remarkably protective, almost tender, in their care of the young. The eggs are laid in shallow pits scraped in the sand of the shore, and the female spends most of the gestation period in the vicinity of the nest. She must leave to feed, however, and during her brief absences the main threat to crocodile numbers, apart from man, strikes. The monitor lizard is uncommonly fond of crocodile eggs, and given the chance can wipe out a whole nest in a very short time. The mother's attentiveness not only discourages monitor lizards, but also means that when the little crocodiles emerge she can escort them across the few dangerous metres from nest to water. Large predatory birds, like hawks and marabou storks, could eliminate a whole season's progeny as they scramble across the open shore. The mother is an effective deterrent and even continues to guard the young during their first few weeks in the water. They also share her meals, snapping up bits that fall from her jaws.

This left-over dinosaur represents the last top carnivore which is a truly integrated part of the aquatic African foodchains. Except for nesting and basking, the crocodile spends its life in the water. The crocodile eats in the water, excretes and dies in the water, in contrast to the alien predators of lake produce, who usually die out of water and bequeath their materials to a terrestrial ecosystem.

# One Prey – Many Predators

The most conspicuous predators of lake produce are the two dozen species of fish-eating birds, all of whom may be dependent on one or, at best, a couple of species of fish. The birds are like fussy eaters who pluck morsels of meat out of the algal vegetable soup. Their main tool for the job is the beak, strikingly similar – long, robust, sharp – in birds as different from one another as the kingfisher and the stork. Methods are varied. Kingfishers dive through the surface, storks wade and stab, cormorants and darters 'fly' under water in pursuit of fish.

In day-to-day life, animals interact with the environment on a one-to-one basis. We see discrete moments in a continual process of predation – one **pied kingfisher** (left) eats one *Tilapia*; a little **malachite kingfisher** (middle) eats a smaller *Tilapia*; one **white stork** (right) catches large *Tilapia*; the **red-legged stilt** (foreground right) stalks another *Tilapia*: several individual fish are sacrificed. Although such interactions are exciting to see and interesting in their own right, it is the sum total of such events on the level of the population that determines the nature of a biological system. Thus these four birds are representatives of four of the many predator populations, which in Lake Nakuru, for example, make a living out of just one *Tilapia* population. The effect on the ecosystem of the one bird–one fish interaction is trivial taken by itself. As we saw in the wooded grassland, the key to the very existence of a particular predator species is the numbers of prey which are available.

A relatively simple ecosystem like the lake, with several predator species depending on just one prey species, is particularly vulnerable to perturbations. If disease, or a fluctuation in the amount of algae, were to drastically reduce the numbers of *Tilapia*, all species of fish-eaters would be out of business with no alternative food source. The species structure of the bird community would change overnight, because the links in the food web between prey and predator were too few.

When we observe several predator species populations feeding from a single prey type, we might think that the predators were necessarily competing. Indeed, overlap on one vital resource is the set stage for competition, but not necessarily the enactment. Competition in this case would only be demonstrated if it were shown, for example, that the kingfishers took enough fish to deprive the storks, and that as a result the number of storks was reduced from what it would have been with no kingfishers around. The individual bird may feel the pinch, but the effect is measured in the population.

# Pouch Fishing

The bird whose beak can in fact hold more than his belly can is the most important African fish eater. Both pelican species found in Africa – the white, which is larger, and the pink-backed, are floating fish nets. The sack of membrane slung between the rims of the lower jaw inflates to an enormous two and a half gallon capacity when dragged under water. How the bird can even swim with it distended and full of water is a mystery. But clearly the system works.

The **pink-backed pelican** fishes alone: presumably it is able to see and strike at particular fish. Sometimes a fish might leap from the water in an attempt to escape (lower right), but inevitably the pelican's net has been cast to sweep up the morsel as it falls back to the water.

The whites are more abundant than the pink-backed at Nakuru, perhaps because of the limited availability of suitable trees for the smaller and more vulnerable birds to roost in. It is the **white pelicans** (inset) that fish in 'flotillas'. A line of two to twenty birds swim along in an oblique formation. At a signal we cannot detect, the line closes smoothly and quickly into a horse-shoe whilst the birds dip into the water in perfect unison. Heads come up, the odd fish will be swallowed by one of the flotilla, the line re-forms and the birds swim on.

In terms of fish caught per individual pelican there seems little difference between the flotilla and solitary methods. A diligent researcher has noted that a lone pink-backed made 14 strikes in ten minutes and caught 8 fish; while 1 white pelican of a group made 86 sweeps in ten minutes and caught 9. It was observed that lone birds fish mostly in the morning when the waters of the lake are clearer and the glare from an overhead sun is absent. As the day wears on, glare increases and the turbidity of the water builds up as algae are stirred by freshening winds. We can only infer that the flotillas drive fish blindly, partially surround them after a tested time, and then dip the nets where the fish should be. It seems reasonable to suppose that when it is impossible to see below the surface, the flotilla method is more successful. Although each flotilla member clearly spends more energy per fish caught, it probably gets more food than it would fishing alone at that time of day. As is often the case, this sort of hypothesis is difficult to test, for it is doubtful if a bird could be forced to hunt singly in turbid water so that we might record its success rate. But this fact in itself could be a tentative demonstration of the hypothesis.

**Carnivores:**
**Pursuit of food**

# Flight

Flight, like the amniotic egg, changed the nature of animal life on earth. Insects took to the air at least 100 million years before the vertebrates, but inevitably there were hungry earthbound creatures that chased after the flying morsels. Among them were the Thecodonts, a group of small two-footed dinosaurs from whom all birds are descended.

In the very earliest stages, vertebrates' 'flight' was little more than a hop, a skip and a jump with the forelimbs outstretched to maintain balance. But over millions of years, the Thecodonts' scales grew longer and the section of the forelimb was thus broadened and thickened until an airfoil was created – the archetypal wing. Jumps lengthened, speed increased, more insects were caught.

The airfoil is the key to all flight. Because the curved upper surface of an airfoil is longer than the flat lower surface, an air current split by the leading edge must flow faster over the top than along the bottom in order to re-combine at the trailing edge. But since it is the same quantity of air – top and bottom – it can only flow faster by becoming thinner. Thinner air weighs less, so consequently there is less pressure on top than below, and the airfoil is literally sucked up into a vacuum of its own making.

Gradually the scales on these incipient wings became feathers, and as the form was further refined, the reptiles were able to glide after their insect prey. The selective advantages were enormous: a whole new world of airborne food, the ability to move from food source to food source, and escape from the rapidly evolving mammalian predators.

The wonder of the bird's wing is that it is both wing and propeller, providing lift to keep the bird up and thrust to move it forward. The inner half of the wing provides the main lift component. The outer half flaps at a greater amplitude and by a twisting motion on the flick of the down-stroke creatures both lift and thrust. A reverse twist and 'feathering' of the flight feathers on the upstroke provides more thrust, but does not reduce the nett gain in lift. The basic pattern varies in speed and twisting to suit the particular species' needs. Sunbirds beat 50 times a second, **pelicans** one and a half; swifts can fly over a hundred miles an hour; sunbirds can fly backwards.

The slotted wing tips of birds which soar, like eagles, vultures and pelicans, reduce drag by changing the effect of the wing on the airflow at the tip. These birds can thus glide with a low rate of sink and with greater manoeuvreability than the long thin-winged albatrosses and sailplanes.

The ability to fly, as we have said, is as inborn as the wings themselves. But the art must be perfected through practice. It is always amusing to watch a colony of young birds like pelicans on their maiden flights. Many of them make their first approaches downwind, hit the ground faster than the legs can run, and end up rolling head over heels.

**Carnivores: Reproduction**

# Mating on the Wing

The **Dragonfly** is an aptly named predator both as an aquatic nymph and an aerial adult. The former looks nothing at all like the latter and specialises in grabbing swimming prey. Large nymphs may even take small fish. At the appropriate moment the nymph crawls out onto a reed stem, splits its skin down the back and steps out a new animal. It takes to the air and spends the rest of its life hawking insects and mating over the lake surface.

The unexpected sight of two dragonflies locked together in the air in an improbable position leads us to wonder how they actually mate. Just before seeking out a female the male injects sperm into a receptacle on the second segment of his elongated abdomen. When he finds a mate he clasps her behind the head with the end of his abdomen. Then, in flight, he drops in front of her, she reaches into his receptacle with the tip of her abdomen and takes up some sperm (upper left).

The function of these aerobatics is obscure. We would guess that, having mated in this way, the pair can then fly over the lake in tandem with four pairs of wings to pull the female out of a near-stall as she dips into the water to lay an egg (lower left).

Dragonflies with 12-inch wingspans were flying about long before the birds evolved. Insect flight is perhaps even more amazing than vertebrate flight since it uses wings which are not modifications of existing appendages. They have developed out of the brittle material which makes up the animals' external skeleton.

The flight mechanism of the dragonfly is relatively primitive. The two sets of wings move independently, controlled by the same muscles which work the legs. Other muscles control the forward component of the wing beats, as well as the twisting necessary not to lose lift on the return strokes. The joints, hinges and pulleys involved in insect flight work so efficiently that they allow a bumblebee to fly, when his wing-loading equations suggest he should not be able to.

Part of the secret of flight in modern insects is a flight muscle which requires no beat-frequency command from the brain. A small mosquito's wings may beat 1,000 times a second, far too fast for nervous impulses to travel down nerves from the brain. The brain message merely switches on the muscle, which then beats at its own inbuilt tonic frequency. Some insects, therefore, have no idea of how fast their wings are moving.

# Cues for the Courting Dance

The cavorting dance of the **yellow-billed stork** is the external sign of the onset of breeding. But what triggers the dance? This simple question would take an entire research programme to answer properly. We can always generalize, though. Behavioural acts are set in motion by preceding events which fall into three timescales.

The first is measured in minutes or seconds. Just before we observed the dance, the external world changed in the eyes of the dancer. He perceived a visual or audio stimulus which 'caused' him to dance. The signal which set the dance in motion probably emanated from a member of the opposite sex.

Secondly, the bird must be somehow ready to send or receive such signals. It must be primed by events which occurred some time before. We now consider a time scale of days or weeks. Breeding seasons are usually linked intimately to ecological events, particularly the availability of food. Somehow the animal receives information from the environment that the optimal time to breed is approaching. It may be a response to a direct improvement in the level of nutrition or to a more subtle signal, like the

change in day length, temperature, even wind direction and humidity. Furthermore, the leaping and bill-clacking of the male may, over the course of the first few days of the breeding season, lead to an alteration in the female chemistry. Whatever the cue, it affects the brain which, in turn, signals the animal's hormonal system. The last glands in the chain of chemical information flow are the sex glands. Their secretions cause physical changes, like brightening of plumage and production of eggs, as well as behavioural changes – nest building, territoriality, courtship and a willingness to mate.

Thirdly, we have the evolutionary timescale, measured in thousands of years. The dance may be tripped by proximal events, such as visual or hormonal signals, but the choreography and indeed the very existence of the dance are products of generations of selective events. Males who danced more attractively, females who responded more readily and discerningly, animals who timed their activities to hatch broods at the best time of year, all were favoured by Natural Selection. These animals produced more young, all of whom received a genetic legacy which included the dance programme.

Carnivores: Reproduction

# The Appropriate Egg

Since one line of small egg-laying reptiles developed into birds during the Jurassic Period, we can state that the egg came before the chicken by about one-hundred million years. Of all major evolutionary events, the amniotic egg is one of the most striking – as we have seen, it allowed the vertebrates to conquer the land and is therefore as powerful in function as it is elegant in form.

As the immediate product of courtship amongst birds and most reptiles, eggs are very precious. They are also obviously organisms in a helpless state. The amniotic membrane and the shell protects them from the physical environment, but what about the animate environment? – particularly predators who relish the concentrated packets of protein and carbohydrates? When the egg's first line of defence – the parent – leaves for reasons of safety or for foraging, most eggs must rely on concealment.

If eggs are secreted in a nest pocket invisible from the outside – a hole in the ground or a tree trunk, or in the foliage of a thick bush, or in a completely enclosed nest like that of a weaver bird – they need no special colour pattern. Equally, eggs protected by formidable parents like ostriches or fisheagles (upper right) can also afford to be conspicuously white.

However, small species like plovers (lower right), which lay their eggs in more open conditions, increase their chances of survival by camouflage. The ground-colour of the eggs matches that of the background and a myriad of flecks and spots disrupts the unmistakeable egg shape. It is relatively easy for an animal's genes to programme a mechanism to produce a systematic pattern, such as alternate stripes or regular polka-dots. But repetitive patterns are more easily learned, remembered and detected by predators. The bird must therefore produce irregularity in its egg patterns. It has to develop a mechanism homologous to a random-number generator in the pigmenting region of its egg-duct. We do not yet know how they achieve this controlled irregularity.

**Carnivores: Reproduction**

# The Inhospitable Nursery

**Pelican** breeding is a sporadic affair, which succeeds only when food is sufficient and suitable sites available. White pelicans are clumsy on land and vulnerable at the nest. They therefore seek out islands, either in the water or surrounded by inhospitable stretches of soda and mud. All these requirements are met only once every few years.

Often the feeding and breeding grounds are not in the same place. The Nakuru pelicans, for instance, breed on Lake Elementaita, where there are no fish. So each day in the breeding season, they spiral up the columns of warming air which rises above Lake Elementaita, glide across the ten miles of intervening hills, and feed at Nakuru. They return in the same leisurely fashion, bringing with them fish for their young. The daily migration costs the pelicans little in energy, since they

use the convective air movements for lift, and barely need to flap their wings. Their infidelity to the lake ecosystem which feeds them, however, costs Lake Nakuru tons of materials which the pelicans leave deposited on the shores and mud flats of Lake Elementaita.

Because of the scarcity of isolated and inaccessible sites, pelicans are forced to put up with crowded conditions in the breeding colony. During the day, the chicks are left in the blazing sun whilst the parents go off to feed. The young have to be remarkably tolerant to heat stress. Their black colouring would seem to be the worst possible choice for sitting on a shadeless mud flat. Perhaps the advantage of being instantly recognised as a defenceless young bird by the crowd of aggressive adults outweighs thermal disadvantages.

When feeding grounds are some distance from the nests, the day's catch of fish is quite advanced in the process of digestion by the time the parents get back. The pouch then serves as a receptacle for regurgitated fish and a trencher for the chicks. Hungrier chicks may even thrust their heads down the parent's gullet and feed directly from the stomach. This devoted feeding technique may have inspired the legend that pelicans strike their breasts so that the young may feed on their blood when food is short: a maternal conceit that adorns the heraldic crest of Oxford's Corpus Christi College.

# The Nest and Territory

The fisheagle is a curious bird to be catching fish. Apart from small spikes on the bottom of its feet to help grasp slippery prey, this specialist is an ordinary-looking black and white eagle. But it has changed its behaviour in striking ways to cope with a plentiful source of food. It still hunts by stooping. However, the instant of strike must be greatly modified. At full stall, with the talons below the water and the extra weight of a snagged fish, the bird has to take off again before it sinks. This requires incredible timing and enormous strength. Perhaps the need for strength explains why it took an eagle, as opposed to a small hawk or falcon, to master the trick.

In general, eagles are relatively rare animals, who pay the price of being top carnivores by having to divide the terrestrial ecosystem into vast territories, each of which can only provide enough food for a pair of birds and their season's offspring. But the fisheagle's switch from relatively scarce and inaccessible land-living food to an abundant aquatic fare allows it to live at high densities – near water, of course.

Along the shore-lines of Kazinga channel in the Ruwenzori National Park in Uganda, **fisheagles** nest every 200 yards, and with each nest goes a portion of the lake – right out to the middle. They spend much more time engaged in territorial display and defence than their terrestrial cousins, simply because their territories are far smaller and they thus encounter their neighbours more often. When not delivering fresh fish to their 1 or 2 chicks, they guard their fishing grounds, either by chasing out other eagles or by uttering their beautiful and mournful call from a shore-line perch.

When the chicks fledge, they spend several weeks sitting around in their parents' territory being fed by both adults. What befalls them thereafter is the fate of many children from a territorial home. The parents' area will not support four adult birds, so the young eventually leave, with considerable encouragement from the adults, when they begin to feed themselves from their parents' fishing grounds. They are literally chased along the shore, from territory to territory, until they end up in less favourable occupied areas with the other young of the year. The strong survive the ordeal, and when an old territory holder dies, its place is immediately taken by one of the young outcasts.

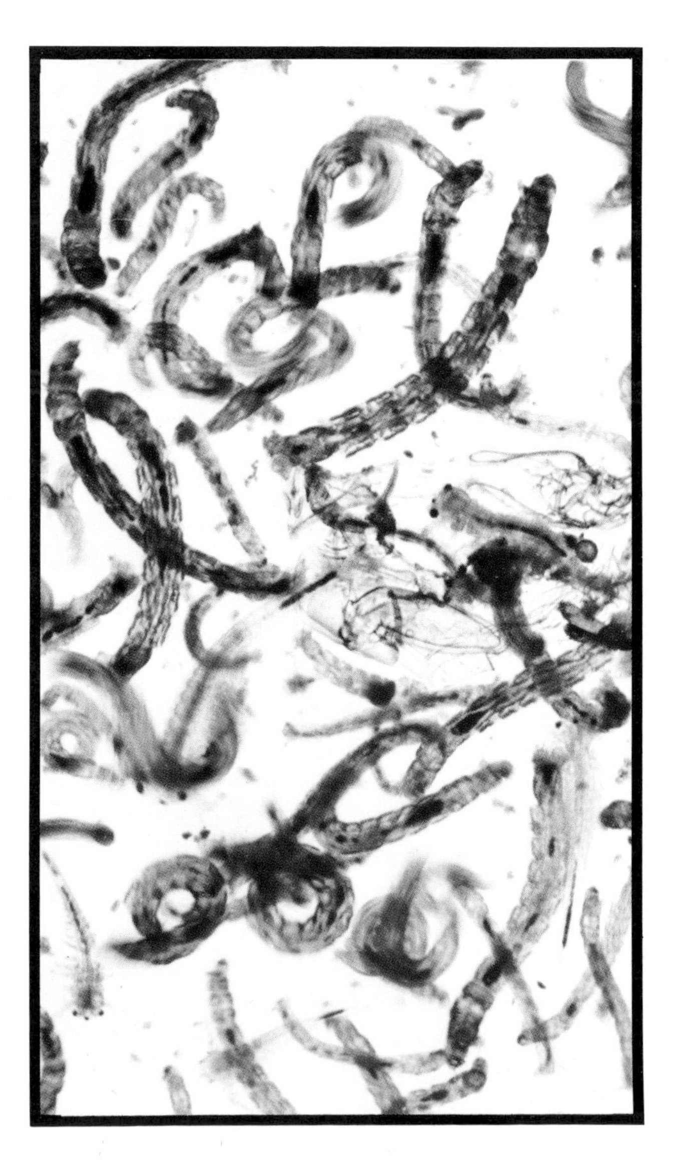

Decomposers: Pursuit of food

# To Eat and be Eaten

In a few drops of water from the bottom of the lake we see minute red threads. They squirm to escape the unaccustomed light. They are **chironomids**, or lakefly larvae (left), who wiggle through the lake-bottom ooze feeding on dead matter which drifts down from the upper layers of the water. Their niche is that of decomposer. In this state their fate is frequently to provide carnivorous fish and beetles with something like 20 tons of meat a day. But the survivors turn into free floating pupae which, when ready, rise to the surface of the lake until, with uncanny timing, the adults in one part of the lake, now transformed into flies, take to the air together (right). From a distance they look like a column of smoke emerging from the water at sunset. Closer, there is a high-pitched whine of wings as thousands of midges whirl about looking for a mate. Once more they become a feast, this time for aerial predators like swifts and flycatchers.

We are frequently puzzled by the 'extravagance of nature', and ask: why lakeflies? What is the purpose of the dense cloud of insects? The question is wrongly put. Where food becomes available, lake detritus for example, some organism will inevitably evolve the wherewithal to eat it. The effect, not the purpose, is the important thing: midges move materials. By using lake offal and providing predators with another source of food, they speed up the flow of materials through the lake ecosystem.

Decomposers: Pursuit of food

# The Adaptable Omnivore

There are some designs which have been used in the animal world for millions of years. They have persisted because they work so well. The outer casings of bugs and beetles, turtles and crustaceans, for example, are elegant in their functional simplicity. Smooth, hard, light and tough, impervious to most enemies, streamlined yet roomy – the domed carapace is an unquestionably beautiful shape. At one of the roots of beauty is an implicit congratulation on surviving thus far.

The **corixid bug** (left) is an aquatic dome-wearer. As a typical 'true bug', its young are nymphs (top left), miniature versions of the adult, unlike beetle larvae which resemble maggots. Corixid bugs are omnivorous, another successful and persisting trait. When primary productivity is low, they can easily switch to a diet of herbivores. When the herbivores succumb to a lack of vegetation, the bugs can feed happily on their dead bodies until the primary productivity recovers.

The protective shell is eventually breached in death. The contents of a tiny **isopod crustacean** (*Lovevla*) spill into the water and mingle with a swirl of microscopic decomposers (right). These are the penultimate step in the aquatic food chain, followed only by the bacteria who return organic molecules and disorganised minerals to the water.

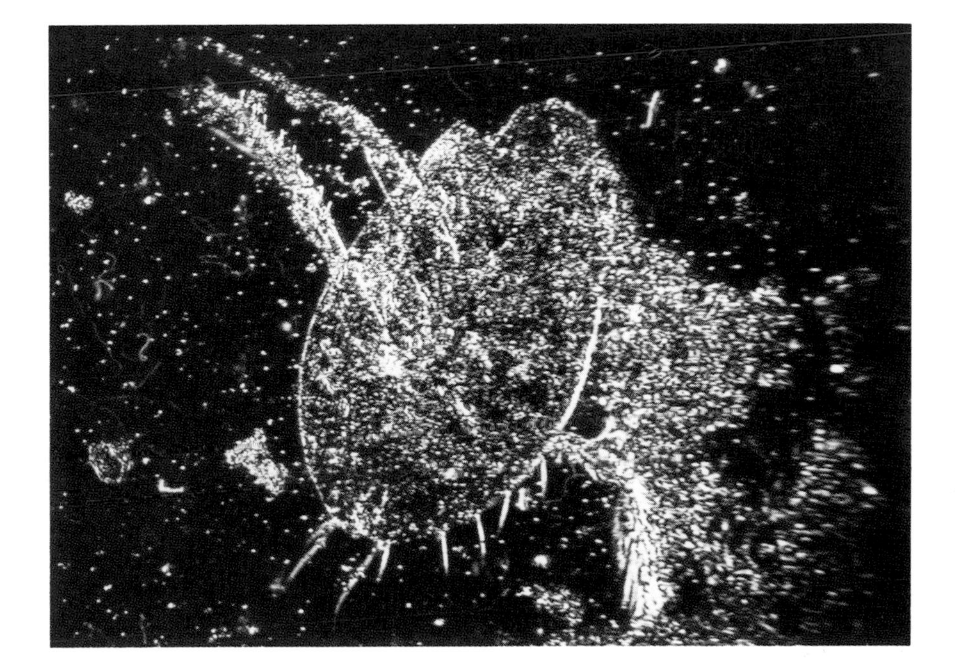

In every ecological system there is the portent of change. The term 'balance of nature' does not describe the state of an ecosystem, it only refers obliquely to the laws which govern it. In our investigations of the natural world we have to freeze its dynamic ebb and flow as a 'momentary stay against confusion'. But no sooner have we fixed the image, than the scene has changed, imperceptibly or dramatically – in nature it really does not matter which. Sooner or later lakes will dry up, but that is no cause for concern, for in their place grasslands will appear.

# III

# FORESTS

# The Essential Forest

Thirty thousand years ago, in a cooler, wetter period, forest covered most of equatorial Africa. The occurrence today of the same species of plants and animals in forests now separated by hundreds of miles of semi-arid bushed grassland is compelling evidence of former continuity. A drier spell from about 25,000 to 12,000 years ago reduced the extent of the East African forests, which also accounts for the paucity of both plant and animal species there compared with tropical forests elsewhere in the world. Not all of the species could survive the dry period.

Once established, a **Forest** contributes to the climate and water conservation that sustains it. Leaves break the fall of rain, and even the most torrential rain drips from leaf to leaf to the floor, seeps through layers of humus and percolates into the soil. Much water is used where it falls and the evaporation of soil moisture is slowed by the canopy awning which keeps out the sun. Moreover, minute particles, exhaled, as it were, from millions upon millions of living and decaying leaves, rise to the upper air layers. Atmospheric water adheres to the particles as ice and eventually falls back to the forest again as rain. It seems then, that forests can even increase a region's rainfall by actually seeding the clouds above them.

The forests' ability to increase rainfall on the one hand and retain it like great sponges on the other, makes them vital to all neighbouring ecosystems. The flow of a grassland spring or the level of a particular lake, indeed the watertable and soil moisture of a whole ecosystem, may be directly linked with the water catchment of the nearest forest.

Forest soils are frequently dark red, acidic from the decay of plant material, incredibly fertile as long as the nutrients used by the trees are recycled. They are, however, unforgiving if abused. Should the forest be indiscriminately cleared, and a season's rain allowed to pound directly on the soil, the soluble nutrients are quickly leached away – the next dry season will produce a brick-hard layer formed by the baked oxides of aluminium and iron. It leaves sterile land that will take a thousand years to remulch.

We have talked of vegetation cycles, but in our time scale the forest appears to be a climax form, an endpoint of tropical plant succession. Certainly only a major climatic change could overcome the stability of the forests, while in the grasslands minor, and quite local effects could turn bushed grassland to woodland in one decade. But a forest is always unmistakably a forest.

Each African forest tends to be composed of a distinct combination of tree species, but even so it would be very difficult – unless we happen to be dedicated botanists – to tell from the bewildering array of species precisely which forest we were in. This is because although they may be different trees, they all look very much alike. For example, the Bugoma forest in Uganda, which is only about five miles in diameter, has some eighty species of canopy trees, all distressingly similar. So much alike in fact are the leaf shape and growth form of most forest trees that they can only be specifically identified by an expert able to reach a flower.

**Primary production: Controlling factors**

# The Path of Water

Life's solvent, water, determines precisely where the forests grow; and the manner in which it is present determines the type of forest formed. There are three main types of forest: **montane,** groundwater and coastal; all are products of local geography and climate. In each case these two factors have combined to increase the amount of water available to plants. If the rainfall is greater than about 40 inches a year, one of the most serious constraints to the growth of large trees is relaxed.

Trace the path of water from the sky, to the mountains, through the land to the sea: in tropical Africa wherever water passes regularly, or otherwise accumulates in the soil you will usually find a forest. On a vegetation map of East Africa, for instance, forests are but a few threads and patches of deep green in amongst the ninety per cent burnt yellow that represents the dry grasslands.

Rainfall potential may be more or less constant throughout the land, but where geographical sculpting has created a mountain to elevate and cool air masses, more rain will fall, and a forest will clothe the mountain slopes – a **montane forest.** Where groundwater is constant all year round, near a spring at the foot of a Rift Valley wall, or along the banks of a perennial river, the moisture requirement of large trees is met, and a groundwater forest results. Or, where wet monsoon winds meet the equatorial coast, heat up, rise and then cool to the point at which they cannot hold their moisture more than 20 miles inland, rain combines with sun to produce an archetypal jungle – the coastal forest.

The forest types, then, are determined by a combination of moisture and altitude, and within the three broad groupings we hear such distinctions as highland, lowland, moist montane, dry montane, moist evergreen, moist semi-deciduous and so on. Although classifying a forest by the kinds of trees growing in it is very difficult, occasionally pure stands of one or two species make classification a bit easier – African olive, cedar, *Podocarpus* (yellow wood), *Brachylaena-Croton*, *Cynometra*, and *Hagenia* for example, are frequently encountered in discrete patches. These specific zones are the result of subtle differences in moisture and altitude within the forest.

# Trees

In the grasslands we continually scan horizontal lines: along the acacia tops, the cheetah's line of chase, the unbroken horizon. In the forest, the axis rotates to the vertical and our eyes are compelled upward. Trees are obviously the dominant feature, their enormous growth engendered by an abundance of water. In the apparent peace of the forest there is a ponderous race upward to reach the light in order to produce what the water in the earth below makes possible. This silent and slow, yet fierce, competition necessitates the structure of the trunk to support a crown far above the forest floor, and, ideally, above its competing neighbours. In the wet of the forest, the growing season is almost continuous and prodigious production goes into the woody structural elements.

This one fact – massive trunks supporting relatively small bunches of leaves – more than any other determines what the forest ecosystem is like. Because so much material is tied up in tree trunks, the atom of phosphorus which passed through the wooded grassland from the plant back to the soil in a few weeks, is likely in the forest to get stuck in a tree for decades. Consequently energy and materials flow much more slowly in the forest than in the grassland. Therefore although the standing crop of the forest is perhaps ten times that of the grassland, the annual production of edible greenstuff is only twice as great – or even less if the grasslands have had a good year.

With the trees hoarding so much material, the pyramid of life in the forest becomes distinctly bottom heavy. waterbuck (left) and other large ungulates do find a niche, but with the bulk of the available forest green matter in the form of leaves above their heads, it is not surprising there are proportionately fewer of them than in the grassland. The majority of the forest animals, herbivores and carnivores alike, are small – insects, birds and small mammals. A large part of the processing of the primary production is left to the decomposers in the soil litter, who receive their nourishment direct from the continual rain of falling canopy leaves.

A typical forest canopy is stratified into three layers. The top layer is composed of

150 foot emergents with buttressed trunks that rise unbranched until out of sight. In the second layer the tree crowns are bushier to make up for the shadows cast by the top layer. In the third, about 30 feet above ground, the crowns are often pyramidal, since sunlight reaches them at an oblique angle, slipping between the taller trees just before and after midday.

Though lower trees get less sun than those in the upper storey, they lose about half as much water from evaporation – a nice balance. But, in the shade on the floor of a mature forest getting enough sunlight becomes crucial: the light intensity is only one hundredth of that which falls on top. Plants on the floor only survive by virtue of a few minutes each day in the form of sunflecks. Many shrubs (right) have evolved broad flat leaves that make the most of these vital moments of pure sunlight.

There are myriads of forest habitats, apparently providing niches of almost endless diversity and richness. But if 50 species of tree occupy one acre of forest, does each have an ecological niche to itself, or are there just a few niches, each shared by an assortment of tree species? This question points up the forest paradox of a diversity of species but a convergence of form: many different types which all look and live alike.

All the trees have similar problems, such as getting enough sun from above ground and nutrients from below, or producing a leaf which is efficient both in catching light and rolling with the punch of a tropical downpour. They live together in an environment which has relatively constant temperature and growth. Thus, evolutionary experiments are many but take the form of species which differ by just a little.

A niche, we have said, is the way as well as the place in which a species makes its living. If two species begin to do the same thing in the same place, one of them should eventually evolve a trick to get more nourishment and reproduce more quickly – to the eventual exclusion of the other. But events between trees occur slowly, and tree reproduction is a lengthy process. Thus not only does it take a very long time for one tree species to 'notice' the effects of another, but also centuries may pass before it is seen to do anything about it. So evolution, or the divergence of form, in forest trees occurs exceedingly slowly; which is why they appear to share the same niche at the same time.

**Primary production: Basic food sources**

# Leaves

The similarity of form in the forest is not limited to the big tree vertical structure. Closer to, we are struck by the similarity of the leaves as well, elongated, tapering, 5 to 10 inches long, dark green with a waxy shining surface – once you have seen one you have seen virtually all of them. Such conservatism can only be born of a common need.

The shape of the archetypal forest leaf is one that spills off heavy rain most effectively without sacrificing size; and the size is most likely a nice compromise between an area large enough to maximise available sunlight, but not so large that it shades the lower leaves of the same tree. The shining leaf surface (left) is produced by wax glands in the leaf epidermis, and apart from providing a sun and waterproof coating that controls evaporation, it may also be a filter to reduce the amount of harmful ultra-violet radiation reaching the food-making cells. Finally, when we look up at the canopy in the morning or afternoon, we notice that many leaves are shining, actually reflecting light downward from part of their curled surface. Perhaps the waxy layer is also a reflective device, designed to share sunlight with the leaves lower down.

Trees, of course, do not eat in the sense that animals do, but like all organisms they need some form of food to sustain their growth. Like all plants they use water, carbon dioxide, nutrients from the soil, and sunlight, to energize the food production factories in their leaves.

The simple equation of photosynthesis – carbon dioxide plus water plus sunlight in the presence of chlorophyll produces oxygen and sugars – is a metaphor for a complex biochemical process. The green pigment chlorophyll is the key. It exists in discrete packages called chloroplasts in green cells.

The impact of light on a chloroplast releases an electron which creates an instant of chemical instability. High energy chemical bonds are shifted, and atoms are borrowed from the molecules of carbon dioxide and water. When equilibrium is re-established in the plant cell a thousandth of a second later, we find a molecule of sugar and one of oxygen – a transmutation worthy of a philosopher's stone.

A current hypothesis suggests that chloroplasts once existed on their own, similar to the numerous forms of bacteria which are capable of making their own food. During the course of green plant evolution, these bacteria-like beasts found a niche in plant tissues and now exist in symbiotic relationship with the plants: the chloroplast enjoys an ideal chemical environment in the leaf, the plant uses the chloroplast as the crucial link in the photosynthesis chain.

Food-making takes place in a layer of cells sandwiched between the upper and lower leaf epidermis. The sugar enters the plant's vascular system (right) to nourish the cells or to be stored as starch. The oxygen not used by the plant is exhaled through trap door cells in the leaf surface.

The rest of the plant is solely designed to get the leaves to the light and allow the plant to live out of water. The trunk, branches and twigs of a tree encase and support the circulatory system. Internal vessels, made of special elongated cells, take water and minerals from the ground to leaves, more peripheral vessels take sugars and waste products away from them.

The food-producing requirements of a forest tree demand that water be pumped from below ground to the leaves, perhaps 100 feet above. But how? The principle force comes, surprisingly, from the top of the tree, where water evaporating from thousands of leaf pores in the process of transpiration, creates an enormous suction in the circulatory system. This negative pressure – 7000 lb per square inch in a top layer tree – literally pulls water up. Water is a remarkably cohesive substance, and each molecule drags another up behind it. The entire tree is affected by this upward pressure, and at midday when transpiration is at its peak, it is actually possible to measure a decrease in the diameter of the trunk as a little water is sucked from each cell.

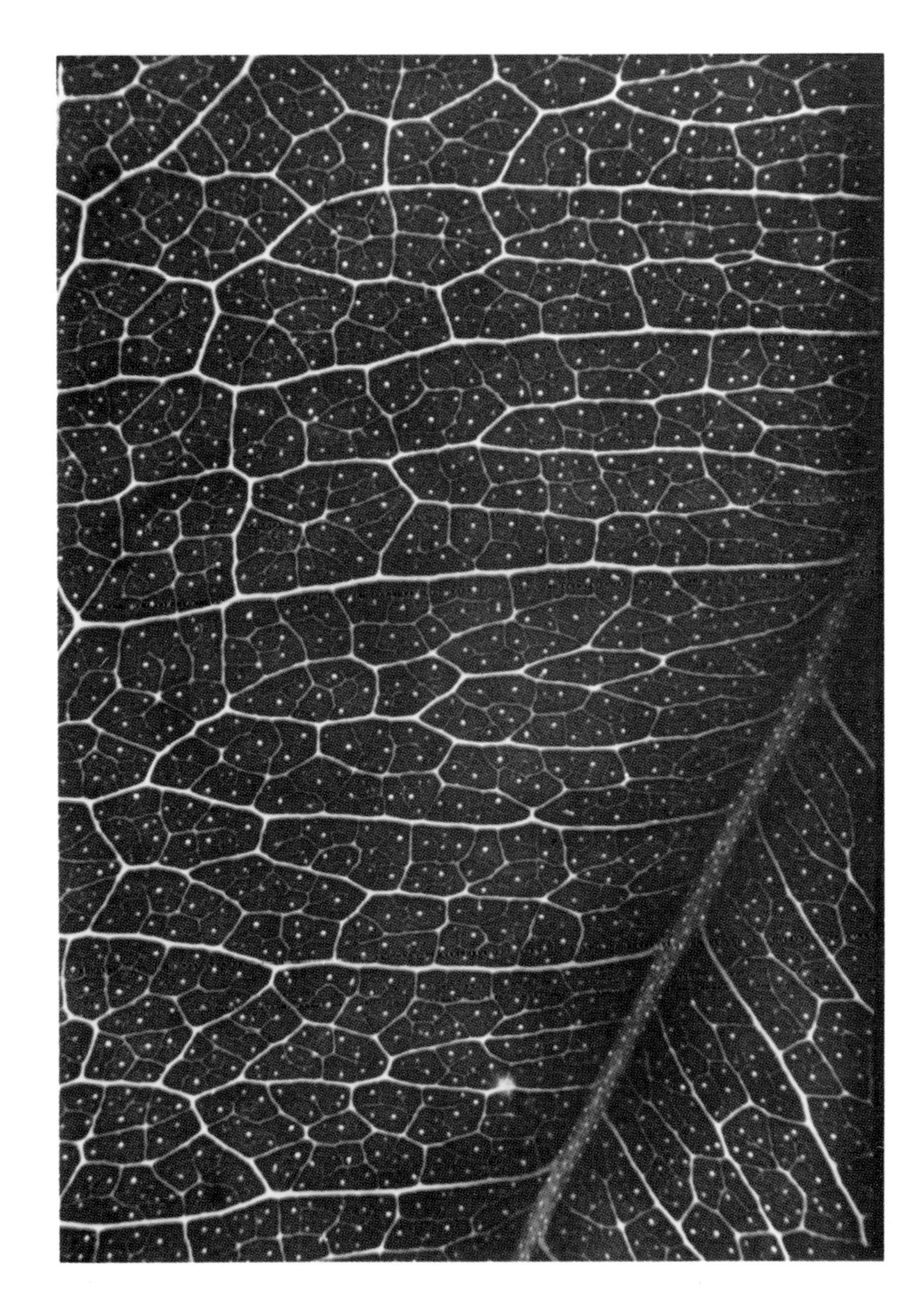

Primary production: Basic food sources

# Giant Grass

The quality of the environment changes so rapidly up on a mountain slope – temperature drops 5° F. every 1,000 feet, moisture increases, the sun is fiercer in the thinner air – that the boundaries of tolerance for various species are abrupt. We step literally from one vegetation zone to another as we climb.

Vegetation changes in the grasslands produce a mosaic effect; on a mountain slope we find belts of vegetation which gird the peaks. One such belt occurs between 7,000 and 10,000 feet on tropical mountains like the Aberdares in Kenya where a long season of almost continuous cloud produces enough moisture to support bamboo. Kilimanjaro is one of the driest mountains in East Africa, being totally exposed to desiccating winds which leave most of their moisture miles behind at the coast. Consequently there is no bamboo zone on Kilimanjaro.

Like many forest plants **bamboo** is gigantic in proportions, with stems forty feet high and four inches wide. This is not so remarkable perhaps, compared with some of the forest trees, but impressive for a a stalk of grass, which is what bamboo actually is: a grass which has established itself as an ubiquitous forest species with an if-you-cannot-beat-them-join-them lifestyle. Perhaps it is an evolutionary response to the strictures which the forest environment places on orthodox grass species.

When we said there is relatively little grass in the forest we were, of course, referring to the grasses that ungulates can eat. Bamboo, though grass and enjoyed by **Sykes Monkeys** (right), elephants, buffaloes and virtually every herbivore that can reach the leaves or find a shoot, does not make an appreciable difference to the size of the large ungulate populations in the forest.

Bamboo grows to maturity, flowers once and then dies – just like a disproportionate annual grass. But what the annual does in one short year can take the bamboo thirty, or even more. Just what triggers the bamboo's flowering and seeding is not known. It is probably not a general response to climatic changes, since different stands in the same forest may have different periods to their flowering calendar. Entire stands flower together, though you must look closely to see the minute one quarter inch long trumpets (lower left), and once flowering is over, the stand dies together, stems split and fall (upper left), leaving room for forest shrubs to proliferate.

**Primary production: Basic food sources**

# Decomposing Plants

The moisture and heat of the forest environment, the decay and slow turnover of forest materials allow a host of lower-class plant citizens to suck a living from the offal of the higher plants. Mosses and liverworts, lichens and fungi all share with the ferns the two-stage sexual-asexual reproductive strategy, the ability to propogate by budding, and the dependence on large amounts of water. They spread their spores and fertilized seeds on the wind, or drop a part of themselves which grows, or simply creep into most of the darker, wetter parts of the forest.

Mosses are the familiar lush cushions of green carpeting damp stones and filling moist crevices. Liverworts, though everywhere in the forest, are hardly ever seen, and if seen, pass unrecognised. They are simpler in structure than mosses but they are also capable of photosynthesis.

The ubiquitous **Spanish moss** festooned over forest trees (middle) in the wetter areas is actually a **lichen**. So are the patches – often circular – of yellow, orange and brown clinging to riverside rocks (right). They are composed of a fungus related to yeast, and an algae. These two very different types of plant live together in a symbiotic relationship. The algae is a single-celled plant capable of photosynthesis and so provides the partnership with food. The fungus provides anchorage, carbon dioxide and water. Lichens are the most widely distributed plants in the world, from the arctic to the desert, and are usually the first pioneers on unclaimed rocks. They start to change rocks to soil. The forest species have a relatively easy life and contribute substantially to the return of materials to the soil.

A **fungus** (left) is devoid of chlorophyll, and so cannot manufacture its own food like the green plants. But undeterred by this fact, the fungi make a living in a number of other ways. Some, as we have seen, are symbiotic. Others are either parasitic or saprophytic – basically the same thing, except the former types derive nourishment from the

juices of living organisms, the latter from dead. Still other forms are predatory, they grow quickly around or into tiny organisms and actually eat them up. Fungi vary in form from the undifferentiated mass of protoplasm called slime moulds to the relatively well-organized mushrooms.

The simplicity, perhaps even inadequacy of our pyramid metaphor is demonstrated very clearly in the forest ecosystem where some plants consume other plants and many are decomposers – where, in fact, 90% of the primary production never reaches the herbivores, the second trophic level. Instead it falls directly to a host of both animal and vegetable decomposers on the forest floor – worms, moulds, fungi and bacteria who thus short circuit the classical food chain. In part, of course, this is the forest's way of feeding itself, for when the decay is complete, the elements are back in the soil again around the roots of the tree that last fixed them in the canopy leaves.

With decomposers in effect performing the functions of both herbivores and carnivores in moving most of the materials along the forest food chain, it is not surprising that as much energy is tied up in them as there is in the more conventional and conspicuous consumers such as large mammals. Of course, the forest floor decomposers do speed up the turnover of materials, but they use a lot of energy in the process. This fact, together with the trees' hoarding of materials in the form of wood, impoverishes the lot of the higher animals in a forest ecosystem.

# Dependence on Others

The apparent serenity of a mature forest belies the fierce and constant struggle between trees for light at the canopy level and for nutrients at the root level. The trees also have to vie for the attention of animals on whom their reproductive success depends. When a plant flowers, the blossom must provide colour or odour to attract, and nectar to reward the animals who act as couriers for fertilization. Plants cannot move, but animals make up for this immobility by transporting the male pollen to the female ovarian receptacle.

The problem does not end with pollination. In most cases ripe seeds would have little chance to grow if they fell into their parents' shade. Their chances of germination and growth are far better if they are dispersed, for then the probability increases that they will land on a temporarily vacant spot, such as a break in the canopy where a tree has fallen. Wind dispersal cannot work as well in the shelter of the forest as it does in the open grasslands. The only other agent is an animal. But animals need to be persuaded or tricked into doing the job, hence the majority of forest trees package their seeds in an edible casing, like a nut or fruit. The massive *Vitex*, the ubiquitous *Chrysophyllum*, the figs and the African olives, all provide monkeys, trogons, hornbills, bats and bushbabies with nourishing treats. The seeds are generally impervious to digestion and are excreted or simply spat out whole, hopefully in a place where competition is less keen.

The seedlings of most forest trees are remarkably tolerant of shade and grow extremely fast, perhaps like the etiolated potato in the cellar. This adaptation demands a drastic difference in the chemistry of the seedlings. When they reach their place in the canopy they settle down to spread and grow as much as their neighbours will let them for the rest of their lives – which may be two human lifetimes.

One way to avoid competition in becoming established on the shaded, root-filled ground is to set seed well above ground. Numerous species of **orchid** (upper left), lichen and most figs drape themselves on other trees. They take water directly from the moisture in the air through specially thin-walled aerial roots.

Figs lure animals to their tasty fruit (lower left). The sticky seeds are passed out in due course or fall from fur or feathers into the crotch of an established tree. The fig seedling begins life benignly enough, just using the host tree for a place to perch. Once established though, the fig roots grow quickly down and around the host's stem. And thus begins an act of parasitism which may last one hundred years. Figs, particularly the **'Strangling Fig'** (right), are relatively fast growing: before very many decades have passed there is more fig than host. Inexorably the host tree is choked to death; finally it succumbs, having given support and its place in the canopy to the fig.

**Primary production: Reproduction**

# Sexual and Asexual

When we see the forest light filtering through the leaves of tree ferns, do we not at once recall that 'the sporophyte of all pteridophytes is, at maturity, independent of the gametophyte'? Perhaps not, for the oddity of fern reproduction usually lies forgotten in our early text books and is overshadowed by the proud advertisement of the flowering plants.

Similar giant fern-like plants, over 30 feet tall, were the first colonizers of the land 500 million years ago when most animal life still squirmed about in the water. The ferns (pteridophytes) are indeed anachronisms, but not out of place in the forest, a realm of vegetative extravagance, giantism, rapid growth, exotic colours and smells. In that earlier time when plants had no animals to rely on as couriers in the reproductive cycle, ferns were dominant. They were successful too, as the world-wide distribution of the common bracken testifies. The prehistoric looking **tree ferns** of Africa persist in forests – a statement about their success and the constancy of the forest environment over time.

Reproduction is a two-staged affair, an alternation between the sexual and the asexual. The undersides of the leaves of the 'adult' fern (the sporophyte, upper left) are lined with special spore-producing capsules which release millions of asexual spores. 'Asexual', because they are viable without the union of male and female elements, and are, therefore, genetically identical to the parent sporophyte. The spores fall on a suitable surface like a moist ball of elephant dung, and grow into tiny 'plants' called gametophytes (lower left). These live on the juices of decay and the effluent of death and hence belong to a class of decomposers called saprophytes. Each gametophyte is a hermaphrodyte; that is, it sports both a male and a female organ. The male part matures first, a safe-guard against self-fertilization, and releases microscopic male cells which literally swim off through the surface moisture. Like a plant version of sperm, they seek out the female parts of another gametophyte, which will have matured by the time they arrive. The fertilized female part then grows into another 'adult' fern. And so it has gone on, as long as there has been moisture enough to provide the medium of transport for the tail-wiggling male cells.

The asexual half of the fern's reproductive effort is a process which allows for no change in form, none of the experimental re-assortment of the characteristics of male and female parents. This mechanical conservatism slows down the possible rate of evolution of the ferns, and, along with a constant forest environment and the apparent distastefulness of ferns, has left us with tree ferns not much different from those the dinosaurs brushed past.

**Primary production: Reproduction**

# The Longest-living Tree?

Like a squat prehistoric monster, the **baobab** tree grows as easily in the coastal forest, as it does in the semi-arid grasslands where we see it here. Its enormous bulk of soft wood holds moisture like a huge sponge and presents a relatively small surface area to the desiccating sun. Legend has it that God, in a fit of anger because the baobab could not decide in which habitat to settle down, threw the tree over His shoulder. It landed on its crown and has grown, roots upward, ever since.

Perhaps just as legendary, although more widely accepted by scientists, is the contention that the baobab attains an age to match its proportions. Large baobabs, which may have a girth of fifty feet (horizon, right), are said to be thousands of years old. Unfortunately the soft wood defies an accurate count of growth rings.

But a characteristic of the tree's growth form suggests an alternate hypothesis to us. Out of the top portion of a mature baobab, we observe two to four main 'turrets' which within a surprisingly short distance subdivide into smaller branches. Follow the lines of the 'turrets' down and they appear to cap major portions of the trunk. Is it possible that several seeds germinate simultaneously out of the five-inch pod, or from the spot where an animal has spat them out? If so, then several seedlings would grow cheek by jowl, eventually touch, and being soft and fast growing, fuse into one disproportionately wide stem? A look at a younger baobab strengthens this idea, for the trunk subdivisions are even more apparent.

The hypothesis, if true, in no way lessens the magnificence of a baobab which, for all the diverse life it supports in its branches, is rather like a one-tree ecosystem. Our wonder should be as great for a 200-year-old tree of such proportions as it is for one 2,000 years old.

# Holes in the Canopy

Suddenly, from the dark of the forest, we stumble into a blaze of light, a large hole in the forest canopy – a clearing. From the air we can see them as light spots regularly spaced. They are usually depressions causing impeded drainage in the otherwise sloping topography. The soils of the clearings are therefore slightly different from the surrounding forest. Frequently there is a waterhole in the centre. This attracts large herbivores such as rhinos who help keep the clearing clear.

Being holes, the clearings have edges called ecotones. These are bands of rapid gradation from the conservatism of the trees in the forest to that of the grass in the glade. The edge-effect is usually one of richness, fast growth, diversity and a high degree of attraction for species endemic to both sides of the edge. Thus the forest meets the grassland not only at the periphery, but deep inside as well.

# Animals from the Grasslands

Because of the presence of open glades, some of the large herbivores of the grasslands are also found in the forests. **Rhino** and **buffalo** and **elephant** all make use of the ecotones and the open patches, as well as the lower vegetation layers within the forest. Their social patterns appear to be more or less the same as their wooded grassland cousins, but their populations and group sizes are smaller. As we have seen in the grasslands, buffaloes and elephants are mainly grazers, so because only a small proportion of the plant biomass in the forest consists of grass, these animals form a relatively insignificant part of the total biomass of forest consumers.

The forest provides retreats and hiding places, but puts certain demands on its inhabitants, which can be seen by comparing forest and grassland populations. The forest buffaloes and elephants are noticeably smaller in stature, perhaps from a combination of less food to go around and the disadvantages of outsized proportions in the tangle of the forest edge. The relatively weak bosses of the buffaloes and the slender tusks of the elephants say much for the protection, from man in particular, which the darkness of the forest offers.

In the wooded grasslands we pointed out the activities and constraints involved in the flow of materials through the system. The same general principles – such as competition and social organisation geared to food supply, predation and anti-predation – also apply to forest species. Rather than labour the general points, we will touch on them, but concentrate more on specifics, and on how some particular problems are solved. To do this, we have to shift perspective a little, to get closer and go upwards.

Herbivores: Pursuit of food

# The Indigenous Inhabitants

Of all large mammals, least is known certainly about the indigenous forest species. Such gaps in our knowledge stem from the very inaccessibility which allows beasts like the **giant forest hog** (left) to thrive. These grow fat and prosper by specialising on the glades' produce and by hiding in the nearby forest. Large family groups graze their way across the year-round grass production of the glades. And that is about all we know of them.

Similarly, the life-style of other forest ungulates like **bongos** (below), bushbucks, sunis and duikers, are largely a scientific mystery.

They are doubly inaccessible in the fastness of the forest by being shy and widely dispersed. They appear to browse mainly on the shrubs of the forest floor. Their numbers are presumably limited by their food supply. The eyes are relatively large for maximum resolution in the reduced forest light. Their pelage is dark and often spotted or streaked with white, which enhances the effect of camouflage on the sun-flecked forest floor.

We might wonder how such apparently frail creatures manage to survive predation. Their appearance is deceptive. They are like coiled springs which at the least disturbance release in a surprisingly violent burst of energy. When alarmed they disappear in a flash and bound into the foliage so quickly that they seem to vanish. With their stoop-backed posture duikers can run straight through the thickest vegetation with unbelievable speed. It is even difficult for a human to restrain a three foot duiker. An attempt to bite the neck of such a struggling packet of muscle could mean a needle-sharp horn in the eye.

The indigenous forest ungulates are impressive creatures, but relatively unimportant in the canopied ecosystem. This is not an excuse for our lack of knowledge, but a deduction from the observation that they account for only a fraction of the materials moving around in a forest.

**Herbivores:
Pursuit of food**

# A Leaf-eating Monkey

The deficit of primary production on the ground and abundance aloft creates a herbivore niche which requires special skills to exploit. Ungulates are earth-bound, and the price as we shall see, is relatively low numbers in the forest. But monkeys are more mobile. They move up and down as quickly and effortlessly as they move back and forth. The **black and white colobus monkey** is a daring and agile climber who feeds on canopy leaves. Perhaps his gentle disposition is connected with his entirely vegetarian diet. His close relative, the vervet of the wooded grasslands, is a successful opportunistic feeder and an upredictable and mischievous beast at close quarters. The colobus travels through canopy pastures by jumping fearful distances, with his long fur flared out like a cape. The fur may actually have an aerodynamic quality which slows the downward speed of long descents. The hand has lost the thumb, presumably a modification which makes hooking onto passing branches easier. The missing thumb gave the colobus its name, for the Greek root of the word means 'mutilated'.

Colobus monkeys live in troops of usually less than 10, led by a male. They move about constantly within one stretch of forest and are probably territorial in order to protect the troop's feeding grounds from the depredations of other groups. The males utter a ferocious, gutteral roar, which is presumably designed to advertise the troop's presence and to intimidate neighbouring troops.

The function of the black and white coat is obscure. Possibly it serves as an unambiguous signal to keep the animals orientated to one another when moving quickly through the canopy. They are so conspicuous, even from an aircraft, that it is difficult to imagine that the colour is camouflage, say to eagles. Of course, the monkeys may behave differently to an eagle than to an aircraft, and immobile in the canopy a black and white animal may be difficult to see.

The colobus has a fermentive gut to break down the cellulose of the green foodstuffs. This causes what seems to be constant indigestion and waves of belches rich with hydrogen sulphide gas. These are offensive to human observers but they are delivered so deliberately and lovingly when facing a troop member that they may serve as a bizarre form of social signal.

With such an apparently rich food supply in the forest canopy, it is rather curious that colobuses are not more abundant. Perhaps the leaves of many tree species are less edible or nourishing than they seem to be. More likely, the high energy requirements of the colobus' vigorous, almost aerial, life requires a disproportionately large food supply.

# Mouths and Numbers

The **bushbuck** feeds on the leaves of shrubs. As we have seen, most of the forest's primary productivity is in the trees above it. So are the bulk of the forest herbivores. In terms of the movement of materials and the flow of energy in the forest, the really important primary consumers are the invertebrates, the little animals without backbones. Most of them are small, inconspicuous, short-lived but with a high metabolism. Their front ends have to be of a very particular design because the edible things in their world are larger than themselves; the **weevil** (left) has to extract nourishment from a leaf which a bushbuck could swallow whole.

We might be surprised that the forest is not grossly over populated by insects, for their potential rate of increase is astonishing. Suppose one female weevil, weighing a tenth of an ounce, produces a conservative twenty offspring. Say one half of these are females, each of which will be mature in a fortnight and produce another ten females . . . The sixteenth generation alone, that is, only the weevils that hatched in the eighth month, would weigh more than the world biomass of tropical forest plants – some ten thousand million tons. Happily, this potential is never realised: it is rare to see an area over-eaten by insects on a large scale in an undisturbed forest. Somehow the number of insects is controlled. How?

The mechanism of population regulation is complex enough to study in a simple system and virtually impossible to tackle fully in the diversity of a forest ecosystem; but there are guidelines. Two general forms of mechanism could be operating, either separately or in concert: control from within the population itself or control from without by some external agent.

Internal regulation can operate through adverse effects which increase as the population size increases. For example, the more animals that are eating a limited food supply, the less there is for each animal to eat and the lower the chance that each gets enough. The point at which the food per insect reaches the danger level must be near the point where the trees are being rapidly defoliated.

Insects have parasites, like virtually every other form of life, so another 'density dependent' process might be a disproportionate increase in the numbers of these parasites as the insect population grows. This kind of control has been observed in temperate climes but has not yet been demonstrated in tropical forests, though we might expect it to operate there too.

External regulation on the other hand, can work through environmental changes which make conditions temporarily unfavourable to the growth of the population. This is a crucial factor in the seasonal grasslands, but unlikely to be very important in the comparatively stable forest environment.

Insects are small and vulnerable targets to a large number of predators, thus predation is a very plausible check on insect population growth. The fact that insects have evolved a startling array of anti-predator adaptations demonstrates that predators are a very real danger to them.

Finally, it is possible that the plants themselves effectively limit the insects' food supply by being distasteful, or even poisonous. Substances have been found in some plants which seem to play a role in the plants' metabolism, toxic wastes which the plant 'deliberately' retains. But then if this were the only check on insect populations, we would expect to find starving insect herbivores, or else a mosaic of healthy distasteful plants and defoliated benign species.

Obviously something keeps the insects in check; exactly what it is we do not know. But it is not avoiding the issue to suggest that probably several regulatory factors working together keep the insects from eating their potential weight in forests.

# Niches

a

b

c

The invertebrates occupy a staggering number and variety of niches in the forest: a whole world of microhabitats of which we are largely ignorant. The number of species is uncounted; many have not even been named; most, like the **Weevil** (right), are probably beetles, since beetles account for about half the 1.2 million species of animals on earth. This fact led the evolutionist, J. B. S. Haldane, when asked by a prelate what characteristics a biologist would attribute to the Creator, to reply: 'an inordinate fondness for beetles'.

The weevil (Curculionidae) and the **Meloid beetle** (b) might bite pieces from a leaf edge; other invertebrates like the **millipede** (Diplopeda – (c)) bore and chew, or pierce and suck. Some eat wood, living or dead. Some, like the **slugs** and snails, scrape the surface cells off plants. Others live on the heads of specific flowers and in eating the petals perform the invaluable function of pollination. Still others eat only pollen; butterflies, and moths such as the Tenuchid moth (*Amata cloracia* – (a)), feed on nectar. The vast majority can fly so any part of any plant from the ground up is fair game to these largely inconspicuous animals.

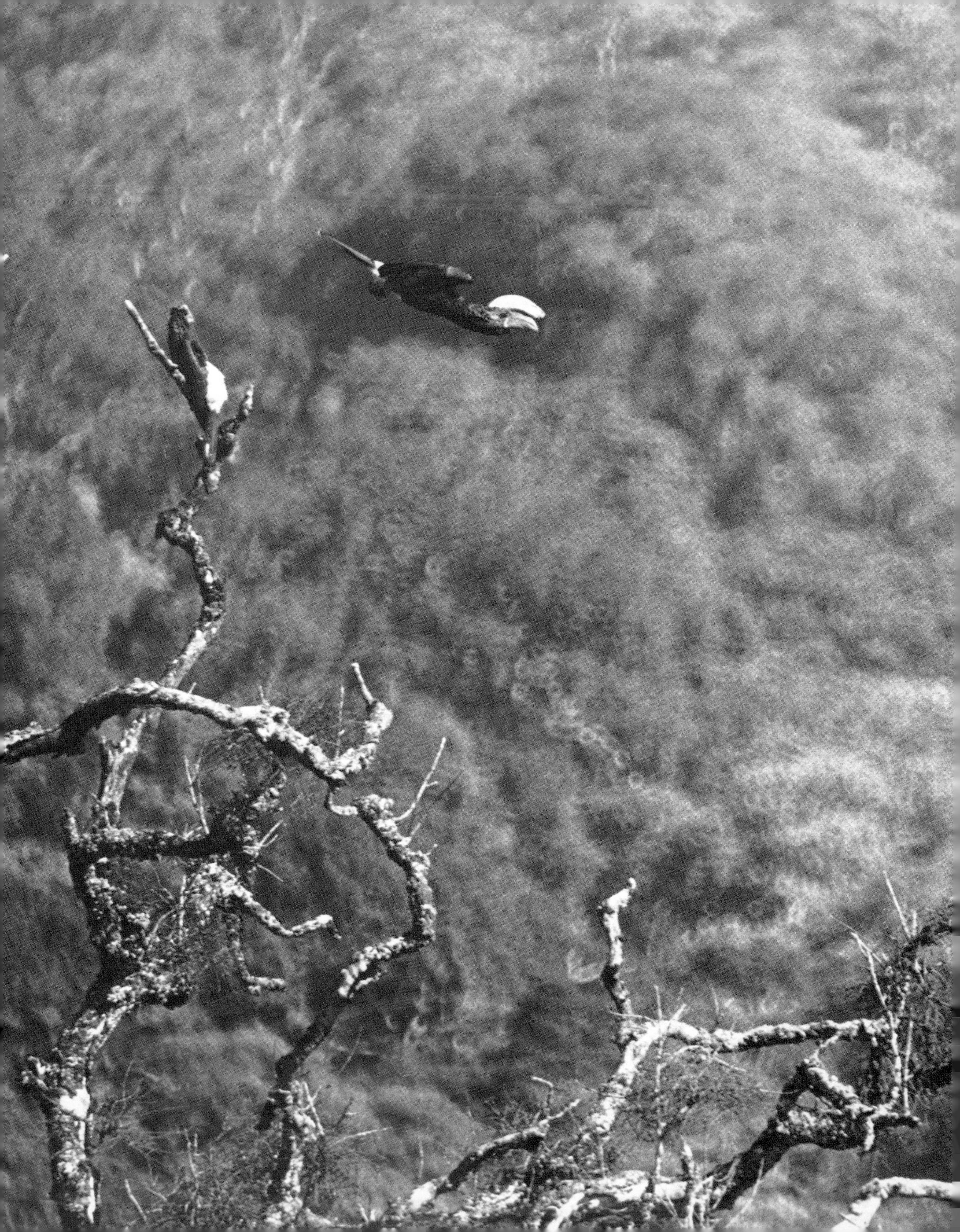

# The Flying Fruit-eater

Birds, of course, seem to us more at home in the tree tops than do mammals. But birds that eat only vegetable matter are in the minority. The high rate of metabolism associated with the physiology of frequent flight is only satisfied by a high-energy, high protein diet. Very few birds eat leaves, and the forest herbivore birds only eat plant produce rich in carbohydrates. Hornbills, trogons and parrots eat fruits and seeds; sunbirds drink floral nectar. The herbivore-plant relationship is advantageous to both forest birds and trees, and both go to adaptive lengths to bridge the trophic gap. We might almost view the trees' response as a pro- as opposed to anti- predation adaptation. Fruits have evolved to be visually attractive to the birds which serve as seed dispersers. Reds, oranges, and yellows are favourite advertising colours. Flowers, too, are gaudily painted in reds, and even shaped and arranged for ease of insertion of the nectar-probe bill of the sunbirds, who drink and pollenate at the same time.

The bills of the fruit-eaters are heavy and sharp-edged for slicing fruit and cracking nuts. The **silvery-cheeked hornbill**'s enormous casque on top of the beak is something of a mystery. It is a honeycomb structure, incredibly light for its size. The casque is certainly used in bill-clacking courtship rituals, and it may also give rigidity and additional strength to the bill for attacking particularly robust fruits.

At the time of year when young are in the nest the food supply may become critical: consequently this is when most of the herbivore bird species jump a trophic level and begin to catch insects and small vertebrates to augment the diet of the growing broods. The birds' ultimate dependence on the reproductive cycles of the trees makes them indispensable to the forest ecosystem on the one hand, but limits their numbers on the other. There is always some fruit or flowers in season in the canopy, but in amounts that are only a fraction of the total plant production.

**Herbivores: Avoiding predators**

# The Poisonous Butterfly

The Monarch butterfly has been poisoned, but with the skill of an alchemist has turned the deed to its own advantage. The food plants of the monarch larva are various members of a poisonous family of 'milkweed' herbs and creepers. To discourage herbivore attacks, the milkweeds have evolved the production and retention of cardiac glycocydes in their veins. The monarch, however, has countered by evolving an immunity to these heart poisons. Since the monarch is the only herbivore known to eat milkweeds (with the exception of one of the ever-present beetles), the plant has not found it necessary to make another move in the evolutionary game.

Not only is the monarch immune to the poison, but it goes so far as to incorporate the chemical into its own tissues, making it a potentially fatal meal for its predators. Once a naive bird has tried a **Monarch caterpillar**, it never forgets how ill it was: in small doses the poison acts as a powerful emetic. Nor does it forget the specially designed colour pattern of the larva, which instantly changes from a bright attractive object to a warningly-coloured one.

Whilst observing the larva, we noticed another trick, When the larva grows large enough to stop scraping the epidermal cells off the surface of the leaf and to start gnawing away at the edge, it begins to cut into the larger veins of the leaf. This releases a stream of the sticky 'milk.' A mouthful of the material causes obvious discomfort, either chemical or mechanical. So, before starting a meal, the larva crawls deliberately to the leaf stem and spends considerable time pinching the stem with its mandibles. This seems to cut off the leaf's circulation without cutting off the leaf. The larva then returns to the leaf proper and consumes the whole thing, veins and all.

# Metamorphosis and Mimicry

The larva has but two functions in the butterfly life cycle – to eat and to grow. It will even eat butterfly eggs, which is why just one or two are laid on a single food plant. Only at night or when it is shedding one of its five skins does it pause. It reaches its full 2 inches in only two weeks, a vulnerable time for a soft worm, even a distasteful one like the monarch larva. It is best to get over this period quickly. We have come to view the phenomenon of complete metamorphosis with the acceptance we afford to a stage conjurer. But consider it again; the larva, a bag of materials, nearly blind, almost senseless, clumsy and thick-legged, sheds its skin a last time and with a wave of the wand is a pupa hanging on the leaf. Inside the pupa the most incredible reorganisation takes place. All parts change size, shape and colour. Mandibles and legs vanish, proboscis and wings appear. Another pass of the wand and a butterfly pops out.

Of course, the larva must go into the metamorphosis with everything set up before hand. The miracle is rigged by the chemistry of the genes. Simple skin cells, which look uniform to us, have messages encoded in their nuclei that they will become wing, antenna or proboscis. Trip the switch with a dose of growth hormones and the cells change and grow, like a beautiful and organised cancer.

The adult has but one function in the butterfly life cycle – to reproduce. It pumps its wings full of the juices bequeathed by its former self and then grows no further. It lives off the fat of the larva and ingests only high calorie nectar to fuel the fires of flying and love. Courtship is airborn and acrobatic. The male sweeps in behind and slightly higher than the cruising female. With fine precision he overtakes her, slows down just above her, drops his tail a fraction, extrudes two bottle-brush like glands from the end of his abdomen, brushes her antennae with them, turns and sweeps back quickly to make another pass. The female is alarmed but overcome by the incredible sweetness of the perfume the male has brushed onto her antennae. This apparently persuades her to land. The male puts down beside her and they mate.

The larva which disappears during the complete metamorphosis not only passes the materials and the blue-print to the butterfly, but, as in the case of the monarch, the cardiac poison as well. The adult is as emetic as the larva; his colours just as striking. Un-

like more furtive tasty species, the monarch flys a lazy cocky pace around the forest edge and through the clearings, as though it knows its warning colour pattern has been learned by most potential predators, mainly birds. Obviously a few monarchs have to be sacrificed to teach the birds a lesson, but the nett benefit to the butterfly population is ninety per cent immunity from predation. If there were more milk-weed, the skies would be orange with monarchs.

So successful is the warning colouration of the **monarch butterfly** (**a**) that other species, tasteful, non-poisonous ones, like **hypolimnus** (**b**) have found it advantageous to mimic the monarch. Even a small patch of orange will make an initiated bird feel uneasy or at least check its attack long enough for the incipient mimic to escape. A survival advantage of only a few per cent is all that is necessary to favour and fix an adaptation in the population. Over many generations, through 'predator selection', which is really a rejection of anything monarch-like, the resemblance is perfected. At this point only an entomologist or a butterfly can distinguish between the monarch and its mimic.

a

b

# Reluctant Pairs

The propensity to live singly and scattered is a common behavioural trait of camouflaged animals. If a predator detects and eats a tasty prey, it looks around the area for more of the same, bearing a 'searching image' of the prey in mind. We, too, know that it is easier to find an object if we have a mental picture of what the object looks like. If the predator finds no other objects which match the searching image, it gives up and looks for something else. If the prey are scattered, the chance is greater that the predator will give up the search before it finds the next prey.

Just before the rains, chameleons forsake their spatial conservatism and begin to wander about. Immobility is an important adjunct to camouflage colouration, but there is no point in avoiding predation if the chance to reproduce is lost. And there is little chance to find a mate by staying in the same bush. So once a year they have to chance detection.

The innate aversion to getting too close to another chameleon is usually overcome by the male's initiative. He pounces on the female and holds her forcibly whilst they mate (Bearded Chameleons, *left*). After the act they quickly part company.

Amongst insects, birds and reptiles, birth takes place most frequently from eggs. But eggs require care or careful hiding to avoid the season's output being eaten at one go. Staying in one place and caring for a nestful of eggs is a potentially dangerous and complicated business, best coped with by animals like birds who are agile and capable of the complex behavioural patterns needed to build, tend and defend a nest.

The bearded chameleon avoids the eggs-in-one-basket dilemma by being one of the few reptiles to bear its young alive. The young chameleons are hidden right through gestation by the mother's remarkable camouflage. The energy cost to the female chameleon is probably not much more than it would be to produce eggs.

The young, twenty six of them in this case, are dropped in a perfunctory way in the space of as many minutes (right). The female takes no notice of them as they wriggle from the amniotic sac which cushions their fall. They are fully developed and will flick their tongue at the first insect they ever see, even before they are completely dry. Each goes its own way immediately, totally independent from the moment of birth.

**Carnivores: Reproduction**

# Parasite in the Nest

There are birds which shirk the parental duty and let others raise their young. Obviously only a few species can enjoy this rather extravagent life-style. The **red-chested cuckoo** is the classical nest parasite. The wandering female watches for a diligent parent like the robinchat to leave the nest, perhaps to feed or to collect some nest material. With the cunning of an oologist the cuckoo has to identify the bird as one who is in the process of nesting and then has to locate the nest. The robinchat boldly chases away the cuckoo if discovered, not because the chat knows about nest parasitism but because it considers most large birds to be potential predators of eggs or young. Given a few unguarded moments, however, the female cuckoo slips in, removes a host bird's egg, lays one of her own and retreats before the chat returns. This in itself is a feat, since the timing of getting the egg out is critical. The cuckoo egg is larger than the robinchat's but this does not bother the sitting bird. Indeed, eggs are extremely compelling objects for broody birds, and large eggs even more so.

The host bird begins incubating. Cuckoo eggs usually hatch first and the chick will immediately set about ejecting the rest of the eggs from the nest. Even if the host and cuckoo eggs hatch more or less simultaneously, the larger cuckoo is much more persistent in begging for food. It eats more, grows faster and is stronger than the host chicks, so it soon succeeds in pushing them over the rim of the nest to oblivion. And then the stage is set for the familiar scene of harassed parents frantically flying about to catch enough insects to feed their idolent and ever-demanding foster chick (left), which ultimately weighs more than both of them put together.

But it's not always like that. No natural mechanism is perfect; success is a quality to be measured in terms of the probable rather than the absolute, and sometimes the cuckoo will fail. In the sequence shown here (right) the cuckoo was late; its egg hatched five days after the host bird's, by which time the larger robinchat was able to demand four times as many meals per day as the parasite. Thus the final infamy of the cuckoo was thwarted; try though it might (lower right) it could not muster strength enough for its innate trick of heaving the host chick from the nest. Perhaps the energy it used in trying cost its life, for it died before the robinchat young was fully fledged.

Carnivores: Pursuit of food

# Hunters in the Air

The majority of forest carnivores are aerial; the most spectacular are birds. Birds of prey are relatively rare in any habitat because of the difficulties in getting at small scurrying mammals from the air. Consequently many eagles are territorial and vigoriously defend their hunting grounds and food supply from intruders (**Tawny Eagles,** left). Specialized bird of prey equipment, such as a large and strong bill, effectively reduces the availability of very small prey, the most numerous type, because their retreats are physically inaccessible. Thus, the majority of the birds in the forest are insectivorous because energy is available to them mainly in the form of insects (**long-tailed shrike,** middle; grassland **flappet lark,** right). Although the proportion of bird species which eat insects is the same in the forests and the grasslands, about sixty per cent, the diversity of species is far greater in the forest. Forests only cover about one per cent of Africa south of the Sahara, yet thirty per cent of the 1,500 odd species of African birds live in forests. This is a direct result of the richness of the insect fauna which, as we have seen, is in turn a function of the richness of forest vegetation.

Catching insects is a full-time occupation even under ideal conditions when they are abundant. Birds use a lot of energy in flying around to catch food for energy. They would seem to be flying in vicious circles if it were not for the fact that part of each meal contributes to body maintenance and reproduction.

The burden on the adult bird is increased several-fold when there are young to feed. They spend the whole day rushing back and forth in a flurry of finding food and stuffing it into the irresistible gaping beaks of the nestlings. During the nesting period an adult bird may lose almost as much weight as the young bird gains.

# A golden-rumped killer

There are few mammalian predators to be seen in the forest. Bats may be heard twittering at night and with luck the golden-rumped elephant shrew may be heard in daytime, rustling through the leaf litter of the coastal forest with its curious long-legged gait, in constant search of its invertebrate food.

He is large for a shrew, about 6 inches long, which is just one reason why he is not a true shrew. No one seems to be quite sure what he is. He may be related to the tree shrews, which are not shrews either, but arboreal insectivores which appear to be the precursors of primates. The golden-rumped elephant shrew may even have fairly close herbivore ancestors, and some workers have ventured the suggestion of an affinity with rabbits. Whatever he may turn out to be for the taxonomists, for our purpose he is a small forest carnivore who spends eighty per cent of the day nosing through fallen leaves catching and eating a weight of insects, worms, millipedes and snails equivalent to his own.

The elephant shrew exhibits the conservatism expected of one so primitive, as well as a unique social system. A male and female share a territory on the forest floor of from 4 to 5 acres. Although they are a pair in the sense that they stay together for several years, in the same home range, they almost entirely ignore each other except for occasional bouts of mating. In these, the female is almost totally indifferent and barely stops searching for food during the act. The male chases off intruding males, the female chases off females, but both tolerate invaders of the opposite sex. The first impression is one of happy domestic territorial co-operation – a misinterpretation of a totally affectionless 'arrangement' which ensures that each member of the pair will get enough food for itself.

Carnivores: Pursuit of food

# Surprise

The wealth of flying insects in the forests puts a premium on hunters which can either take food in the air, or quickly in the brief instants when the insects alight. The **chameleon's** tongue clearly takes the insect completely by surprise. The alternate tactic for the chameleon would be to rush and snap. But insects are fast and have good eyesight. A hungry beast the size of a chameleon would have a small chance of sneaking up on its prey. So the chameleon forsakes speed of movement. It moves slowly along branches with a quaint rocking motion that has the rhythm of a leaf swaying in the wind. As it moves, special pigment cells – black and green mainly – just under its skin, expand and contract to produce effects which imitate the colours which the eyes see and the brain interprets. The turreted eyes swivel about independently taking in colour and looking for prey in an almost total sphere of vision. When one eye sights a prey item, the other sweeps parallel to fix the object in a binocular stare. Prey suitability, distance and angle are judged in an instant. A small bone flicks the tongue clear of the jaws (top) and with an hydraulic surge it inflates forward, like a child's party favour. Pressure is released, the tongue snaps back (bottom) and the meal is finished in less than a second.

Carnivores: Pursuit of food

# Speed and Precision

The precision of the motor control and complex design of our body, sensory and nervous systems are taken for granted by us because we live by them. Consider a forest hunter like the **praying mantis** (Mantidae). It catches insects by snatching them out of the air, a feat at which we have only moderate success. The mantis must receive instantaneous information about the insect's position, distance and speed relative to that of itself. It must process the information, come to a decision, and strike with its elongated forelegs before the insect is gone. Large compound eyes set wide on the head identify that the object is prey and provide the optical data necessary to feed into the distance equation. Unlike most invertebrates, the mantis' head can turn on its thorax. The tilt of the head excites sensory hairs which signal information about the angle of the prey away from the axis of the mantis' body. The forelegs are cued to the same angle and fired off before the distance changes to a useless value.

It is not surprising if this description uses a jargon echoing that of the control system of a computer operation, for the simile is not far off what must really happen in the animal. The hardware of body and nervous system is designed by natural selection and programmed with a set of genetic code instructions that allow the mantis to catch an insect the first time it tries. But it is not a flexible system; the mantis cannot do much else as well as it catches insects.

Carnivores: Pursuit of food

# The Trapper

Some carnivores go out and hunt their food, others set traps for it. **Sheet-web spiders** build horizontal nets, trampolines of silk, and then wait for insects to fly or fall into them. These spiders are nocturnal for two good reasons – in daylight prey would see the closely woven sheets, and larger predators would see the waiting spider. At the least untoward vibration, which the spider feels with its feet, it rushes out and pounces on the object which caused the disturbance. The chase and kill which follow the moth's arrival is as dramatic as any lion kill. The moth is larger than the predator but at a disadvantage in the tangle of the web. The spider feints and rushes at the thrashing prey trying to inject an immobilising poison and loop some more strands of silk around the moth. If the moth does not manage to struggle free it is wrapped up and set upon immediately, with the spider quietly sucking out its body fluids.

Because the forest carnivores are mainly small, most of them crawl a thin line between the eater and the eaten. It has never been measured, but we would guess that forest carnivores eat almost as many of each other as they do of herbivores. The shrike eats the chameleon who ate the mantis, who ate the spider, who ate the herbivore moth. This sort of horizontal food chain contributes to the conservatism of the forest by slowing down the trophic flow of materials through the ecosystem.

Carnivores: Pursuit of food

# The Hunter

The **hunting spider**'s technique of getting food is less passive than its cousin's sheet web trap. It prowls about the forest floor at night looking for prey with a bank of half a dozen sensitive eyes. These eyes are relatively simple in structure and so several of them are needed to get enough information for an accurate charge in dim light.

Even such formidable beasts are at a temporary disadvantage during the reproductive period. The drain on energy and vitality is severe when producing offspring. The act of giving birth, egg laying or nest building slows down an animal's movement, fixes it momentarily in space and increases an enemy's chance of making a successful strike. But the effort and risk are obviously unavoidable. The female Lycosid spider encumbers herself for an appreciable portion of her lifetime, lovingly carrying her egg mass under her abdomen (below). In this way she can carry on her mobile strategy of hunting whilst protecting her egg case. When threatened, she tucks it up under her, and anyone who wishes to take it will have to deal with her first.

Carnivores:
Pursuit of food

## An Army of Hunters

The most effective and thorough carnivore hunter in the forest is a colony of **siafu** (*Dorylus*), the so-called army ant. Before the rains begin, a dreadful horde of hundreds of thousands of ants pours out of the ground, presumably from the nest centre. It is not known what triggers this march nor what determines the direction. The function is unmistakeably to get food. Any animal of any size, living or freshly dead, which is in the ants' path is quickly set upon, pinned down, sliced up and carried off in tiny pieces back to the nest. Even elephants have been driven to madness by a trunkful of siafu. Soon there is a two-way traffic along the trail, with workers scurrying between rows of alert soldiers – thousands of individuals who are individually worthless, expendable. Soldiers attack against absurd odds: we pluck them off our legs leaving the heads biting us. It is perhaps the mechanical nature of their predation, their totally unemotional, unreasoning, unqualified instinctive life-style that humans find disturbing.

The movements are so quick; all are running or so it seems, translated into human perspective relative to body size. Ants meet head on, pause to touch antennae in instant identification, then rush on. Logs are overrun without pause, ants cling to-

gether to form bridges over small streams in a twinkling, prey items such as grasshoppers (lower right) are set upon and dispatched with assembly-line efficiency. Why the hurry? Perhaps the question is too human to apply to ants. We tend to judge events by our own criteria, and therein lies the 'in-human' quality we ascribe to small animals like siafu. One foraging hunt of these ants may take a worker's life time. That is a long time, to an ant.

### Carnivores: Avoiding predators

# Ancient Messages?

The arrangement of the carnivore's front end, as we have seen, is to expedite dealing with mobile meat. The **serval cat's** threat display (top right) is not displayed at its prey – there is no time for that – but at other cats or larger animals threatening it: a reminder that teeth could be used for fighting or self-defence if a critical distance is breached. The evolution of animal displays invariably makes use of existing body parts. If those parts enhance the message of the display so much the better.

We recoil from the serval's threat or from a column of siafu, yet we reach out to touch a butterfly. Is there an absolute quality about the aspects of herbivores and carnivores that leads us to consider the duiker, colobus or butterfly beautiful and the hunting spider, **leopard** (left) or **white-lipped herald** snake (bottom right) fearsome? Or does our response to members of these different trophic levels stem from a time when we were functionally closer to them – when the herbivore was vital to us as food and the carnivore potentially fatal, or, at best, painful?

The qualities of the attractive and the fearsome, or of beauty and ugliness have been described by poets, but the meanings of the words may have roots reaching deep into our origins and biological selves. One of our unique qualities is the ability simultaneously to experience the thrill, and to analyse the components, of the natural world. Is this inherited from our distant forefathers, whose survival must have depended to a large extent upon their learning to pause and apply a measure of classical, analytical reasoning at times when the animal in them urged uncontrolled flight, or unbridled attack?

We come, inevitably, back to the decomposers; to one which is amongst the most successful organisms on earth – the ubiquitous fly. Among the tens of thousands of **fly** species, such as the Calliphorid, we find all sorts – herbivores and carnivores, predators and parasites – but the vast majority thrive because of death.

It is curious that our reaction to decomposers is one of revulsion, for the world would indeed be a revolting place without them: it would be piled to the sky with dead organisms. They would not smell, since without bacteria there would be no putrefaction: but it would be terribly still, lifeless, ecosystems having run out of fuel.

With the concerted efforts of millions of decomposers from vultures to flies to bacteria, an entire elephant corpse, save the bones, will be returned to the soil in a few months. Thus the ecosystem's store of exchangeable resources is replenished without relying on the long-term contribution of dissolving mountains. The decomposers, then, are not only the link between the organic and the inorganic, they are also the bridge between death and life, which makes new life possible. Next time you kill a fly remember that, in the end, he and the earth will get their own back.

# Epilogue

We have seen that energy and materials flow through all ecosystems along similar routes. Their amounts determine the number of animal and plant pipelines up through the pyramid of life and back down to the soil. If we export materials indefinitely, we observe decreases in the variety of species, in the rate of flow round the system, in the very tempo of life. Every farmer knows this when he finds that the yield of a much-harvested field begins to drop. His solution is to fertilise, to import nutrients into the little system of his field.

The loss from all Earth's ecosystems is eventually to the sea. The sea is the terminus of rivers which carry nutrients leached from the habitats they drain, and the effluents of human use. Under normal conditions, the loss of materials can be offset by the weathering of parent rocks. Under excessive harvesting, however, the rate of flushing enriches the sea bottom, but impoverishes the land.

We can preserve the richness and yield of our terrestrial systems by either waiting for the sea bottoms to rise again as mountains so erosion can begin anew; or by fertilising with materials dredged from the sea; or by using the pyramid of life with the art and sensitivity of plants and animals. We have no time for the first; the second is very expensive.

The third requires knowledge and organization of this exquisite world. As we reason and compare, we must neither lose our sense of wonder, nor be discouraged by the notion that natural beauty may no longer be its own excuse for being.

> 'Unorganiz'd Innocence: An Impossibility.
> Innocence dwells with Wisdom, but never with Ignorance.'

William Blake, Notes on the Four Zoas

# Acknowledgements

Two sorts of enthusiasm made this book possible. One type was evidenced by our editor, Adrian House, and the design editor, Ron Clark. Their enormous interest and the hours they spent on the book are reflected on every page.

The other sort is that inherent in our East African friends and colleagues – naturalists, conservationists, wildlife managers, scientists and administrators – whose work over the last two decades in or on behalf of African ecosystems has produced the majority of facts and ideas we have used. We have not cited individual references, and hope our scientific colleagues will forgive our use of their material as well as any misinterpretations and simplifications.

In particular, we wish to thank the following organizations for their help:

Department of Zoology, University of Nairobi
International Centre of Insect Physiology and Ecology, Nairobi
Kenya National Parks
Nakuru Wildlife Trust
National Museums of Kenya
Ngorongoro Conservation Authority
Uganda Institute of Ecology
Uganda National Parks

and also the following individuals:

David Babu
Edward Bagley
O. H. Bruinsma
Tony Carn
Anselm Croze
Nani Croze
Eric Edroma
Ted Goss
C. J. Heather
Dawid van Heerden
John Hopcraft
Lew Hurxthal
Ian Joyce
Joe Kioko
Mary Leakey
Richard Leakey
Mr. Mburugu
Mr. Mgina
Colin Middleditch
Mulji Modha
P. J. Mshanga
Peter Ndegwa
Arthur Newton
Daphne Nightingale
Ted Norris
Perez Olindo
N. B. Owino
B. Patel
Gaylen Rathbun
Soran Reader
Gerry Rilling
Ginger Schwan
Tom Schwan
David Sheldrick
Johnson Sumba
Simon Trevor
Chris Tuite
The late Desmond Vesey-Fitzgerald
Sergeant Waithwa
Sam Weller
Jonah Western
Bill Woodley

Michael Clifton and Alec Duff Mackay identified small beasts for us; parts of the manuscript were skilfully de-coded and typed by Lynne Colgrave, Nicola Brown, Brigitta Reader and Geraldine Harney; and the whole work was read by Michael Gwynne and Michael Norton-Griffiths, who were gentle but firm in their criticisms. We are grateful to all.

# Suggested Further Reading


*East African Lakes and Mountains*, L. H. Brown (E. A. Publishing, Nairobi, 19 )

*Looking at Animals*, H. B. Cott (Collins, 1975)

*The Selfish Gene*, R. Dawkin (Oxford University Press, 1976)

*Among the Elephants*, I. and O. Douglas-Hamilton (Collins, 1975)

*The Spotted Hyaena*, H. Kruuk (Chicago University Press, 1972)

*East African Vegetation*, E. M. Lind and M. E. S. Morrison (Longmans, 1974)

*Portraits in the Wild*, C. J. Moss (Houghton Mifflin, Boston, 1975)

*Serengeti: Kingdom of Predators*, G. Schaller (Collins, 1973)

*The Web of Life*, J. H. Storer (Devin-Adair, New York, 1953)

*Social Behaviour in Animals*, N. Tinbergen (Methuen, 1954)

*East African Grasslands*, D. Vesey-Fitzgerald (E. A. Publishing, Nairobi, 1973)

*Sociobiology*, E. O. Wilson (Harvard University Press, 1975)

*The Long African Day*, N. Myers (Cassell, Collier, Macmillan, New York, 1972)


# Index